Richard Channing Moore Page

A Handbook of Physical Diagnosis of Diseases of the Organs of

Respiration and Heart and of Aortic Aneurism

Richard Channing Moore Page

A Handbook of Physical Diagnosis of Diseases of the Organs of Respiration and Heart and of Aortic Aneurism

ISBN/EAN: 9783337812416

Printed in Europe, USA, Canada, Australia, Japan

Cover: Foto ©berggeist007 / pixelio.de

More available books at **www.hansebooks.com**

A HANDBOOK

OF

PHYSICAL DIAGNOSIS

OF

Diseases of the Organs of Respiration and
Heart, and of Aortic Aneurism

BY

R. C. M. PAGE, M.D.,

Professor of General Medicine and Diseases of the Chest in the New York Polyclinic;
Visiting Physician to St. Elizabeth's Hospital, the Polyclinic Hospital, and the
North Western Dispensary, Department of Diseases of the Heart and
Lungs; Member of the New York Academy of Medicine; Hono-
rary Vice-President of the Congress held in Paris, 1888,
for the Study of Tuberculosis, etc., etc.

Third Edition

NEW YORK:

J. H. VAIL & CO., 21 ASTOR PLACE

1891

To

ALFRED L. LOOMIS, M.D., LL.D.,

WHOSE THOROUGH AND SYSTEMATIC TEACHING,

AS WELL AS MANY ACTS OF KINDNESS, ARE

GRATEFULLY REMEMBERED,

This Modest Little Volume

IS RESPECTFULLY DEDICATED BY

THE AUTHOR.

PREFACE.

In compliance with the requests of students this volume is now placed before the medical profession. In it I have endeavored to treat the important subject of Physical Diagnosis from a logical standpoint,—the deductions in each case being drawn chiefly from personal observation. By this means I have, in many instances, furnished the all-important missing links that necessarily occur in a mere printed list of physical signs, however ingeniously arranged. The student is thus saved much valuable time that would otherwise be lost in attempting to supply those links unaided. Besides a consideration of the physiology and normal anatomy of the organs involved, brief mention of etiology and pathology has been found necessary in many cases, as well as the proper classification of disease.

There is but little original to be offered at present on the subject of Physical Diagnosis, but whenever I have thought that I had cause to differ with even the most eminent writers on this subject, I have not hesitated, with due respect to them, to do so.

To Dr. Henry Macdonald, 151 East 31st Street, New

York City, I am greatly indebted for the illustrations, most of which are original and copied from life, or pathological specimens.

R. C. M. PAGE, M.D.

31 WEST 33RD ST.. NEW YORK CITY
May, 1889.

CONTENTS.

PHYSICAL DIAGNOSIS.

CHAPTER I.

THE CHEST IN HEALTH.

Physical Diagnosis is the art of distinguishing health from disease, and one disease from another by means of the physical signs presented in each case. It approaches nearer to an exact science than any other branch, and may truly be termed the mathematics of medicine. It embraces various methods of examination, including the use of instruments to be hereafter described.

The Physical Signs of health, as well as of disease, are those that are to be recognized by the examiner's special senses, particularly sight, touch and hearing. There are certain physical signs characteristic of health and others that belong respectively to individual diseases.

In order to understand the physical signs of disease it is evidently necessary first to know them in health. Then, by the application of principles of well-known physical laws, a logical and correct conclusion may be arrived at in each case, which is more reasonable than

1

to attempt to commit isolated facts to memory, only to be forgotten.

For the sake of convenience the chest walls are marked out into different regions, the limits of which, though arbitrary, should always be made with due regard to the anatomy of the underlying thoracic organs. There are an anterior, posterior, and two lateral regions of the chest, and each of these is subdivided into other regions.

Anterior Region.—This is divided into two similar parts, right and left, and a middle part.

The right and left parts comprise, each from above down, the following regions: 1, supra-clavicular; 2, clavicular; 3, sub-clavicular (infra-clavicular, supra-mammary); 4, mammary, and 5, sub-mammary (infra-mammary, hypochondriac). The middle part is divided into the, 6, supra-sternal; 7, superior (upper) sternal, and 8, inferior (lower) sternal regions.

The supra-clavicular region is triangular in shape and is situated above the clavicle. It is bounded below by the upper border of the clavicle, within by the lower portion of the sterno-mastoid muscle, and without by a line drawn from the inner end of the outer fourth of the clavicle to a point on the sterno-mastoid muscle corresponding with the upper ring of the trachea.

On both sides the apices of the lungs rise into the neck above the sternal ends of the clavicles, according to Gray, about an inch and a half, but in persons with long necks as much as two inches; in women rather higher than in men, and on the right side than the left.

The clavicular region corresponds to the inner three-fourths of the clavicle.

The sub-clavicular region (also called infra-clavicu-

lar, or supra-mammary) is bounded within by the edge
of the sternum (sternal line), without by a line let fall
perpendicular from the inner end of the outer fourth of
the clavicle, and continuous with the anterior axillary
line; above by the lower border of the clavicle, and
below by the upper border of the third costal cartilages
and ribs, corresponding exactly with the base of the
heart. In order to find the upper border of the third

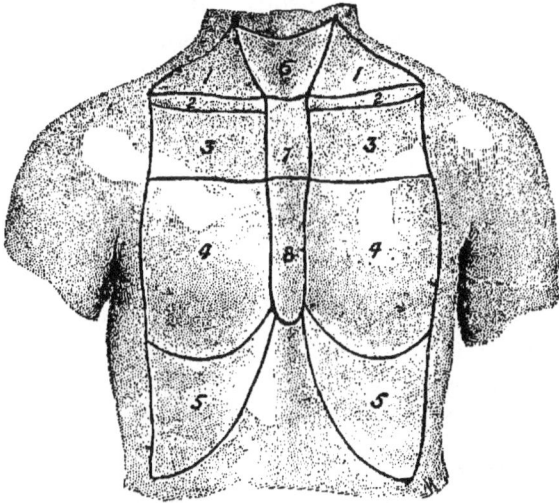

FIG. 1.—1. Supra-clavicular Region; 2. Clavicular; 3. Sub-clavicular; 4. Mammary;
5. Sub-mammary; 6. Supra-sternal; 7. Superior Sternal; 8. Inferior Sternal Region.

rib, especially in fat people, feel for the horizontal
ridge on the sternum that marks the line of union between the manubrium and gladiolus. At this point,
on either side, is the articulation of the second costal
cartilage with the sternum. Immediately below is the
depression between the second and third ribs, or the
second intercostal space, the upper border of the third
rib, as well as lower border of the second being distinctly felt. The right and left second intercostal

spaces, called respectively aortic and pulmonary, have special significance in the study of the heart, as we shall see. The sub-clavicular regions are chiefly occupied by lung tissue, but the right primitive bronchial tube, larger than the left, more superficially situated anteriorly, and given off higher up, causes important differences in the physical signs of the two regions, as will be fully described. The study of these two regions in health is of the first importance, the more so as tubercular pulmonary consumption usually manifests itself first in one or the other.

The mammary region is bounded above by the upper border of the third rib, below by the upper border of the sixth rib, within by the edge of the sternum (also called the sternal line), and without by the anterior axillary line which is continuous with the outer boundary of the region above. The heart is chiefly situated in the left mammary region, the apex-beat corresponding to a point between the fifth and sixth ribs, one inch and a half below, and half an inch within the left nipple. Gray and others, however, place it two inches below, and one inch within the left nipple. The superficial area of cardiac dullness lies almost wholly within the left mammary region. The right mammary region extends down to the liver, the lower border of the former exactly corresponding with the upper border of the latter.

The sub-mammary region (infra-mammary, hypochondriac) is bounded above by the upper border of the sixth rib, below by the free margin of the ribs; within it comes almost to a point at the edge of the sternum and without it is bounded by the anterior

axillary line. The region on the right side is occupied
by the right lobe of the liver. On the left side, we
have the left lobe of the liver and the large end of the
stomach. The outer boundary of the region on the left
side corresponds to the anterior border of the spleen
from the ninth to the eleventh ribs. Between these
two regions is the epigastrium.

The supra-sternal region lies above the supra-sternal
notch and between the supra-clavicular regions. In
it lies the trachea, but by firm pressure downward
with the finger, the patient's head being inclined forward, pulsations of the transverse portion of the arch
of the aorta may be felt, especially in the case of aneurism.

The superior sternal region (upper sternal) corresponds to that portion of the sternum above the line of
the upper border of the third ribs.

The inferior sternal region (lower sternal) corresponds
to that part of the sternum below that line.

Posterior Region.—This is divided on each side,
from above down, into the, 1, supra-scapular; 2, scapular, and 3, sub-scapular (infra-scapular) regions, and
between the scapulae is, 4, the inter-scapular region.

The supra-scapular region corresponds to the supra-
spinous fossa of the scapula, and is occupied by lung
tissue.

The scapular region corresponds to the infra-spi-
nous fossa of the scapula, and is also occupied by lung
tissue. It is much larger than the former, and extends,
according to Gray, down to the eighth rib.

The inter-scapular region is situated between the
scapulae on both sides of the spinal column, which di-

vides it into the inter-scapular regions of the right and
left sides. It extends downward to a line drawn horizontally from the inferior angle of one scapula to the
other. In front and on each side of the spinal column
in this region the bronchi enter the lungs, the right
bronchus opposite the fourth dorsal vertebra, accord-

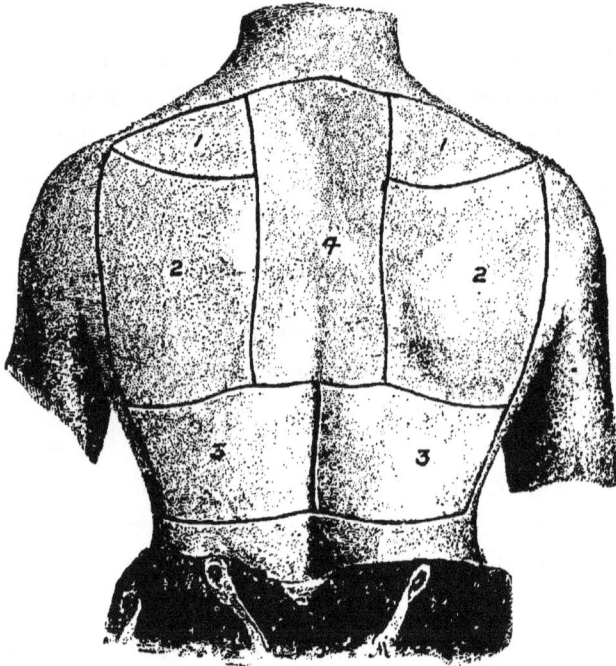

FIG. 2.—1. Supra-scapular Region; 2. Scapular; 3. Sub-scapular; 4. Inter-scapular
Region.

ing to Gray, and left opposite the fifth, about an inch
lower.

The sub-scapular (infra-scapular) region is bounded
above by the lower borders of the scapular and inter-
scapular regions, below by the lower border of the
twelfth rib, within by the spinal column and outside
by the posterior axillary line.

Lateral Regions.—These are divided, right and left, into, 1, the axillary and, 2, sub-axillary (infra-axillary) regions. The axillary region corresponds to the axilla.

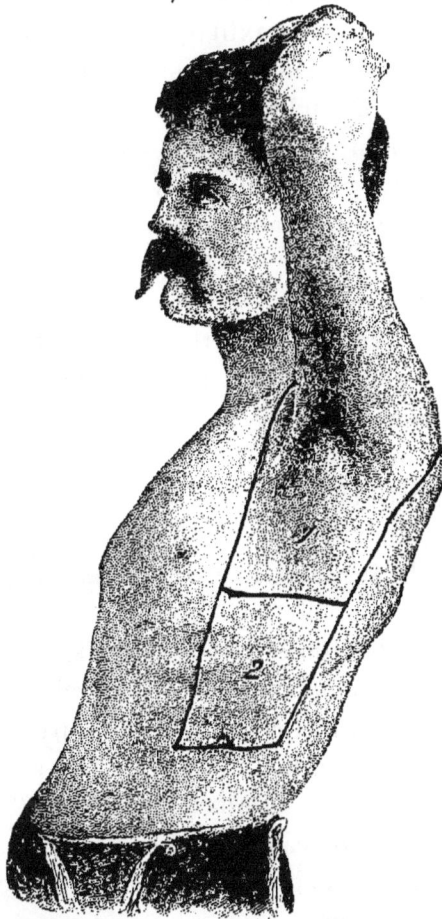

FIG. 3.—1. Axillary Region ; 2. Sub-axillary Region.

It is bounded below by a line connecting the lower border of the mammary region with the lower border of the scapular region; in front by the anterior axillary line and behind by the posterior axillary line. They are both occupied by lung substance.

The sub-axillary (infra-axillary) region is bounded below by the lower border of the twelfth rib, above by the lower border of the axillary region, in front by the anterior axillary line and behind by the posterior axillary line. On the right side is the liver, on the left is the spleen and large end of the stomach.

A line drawn perpendicularly downward from the middle of the axilla is called the middle axillary line, or simply the axillary line. It is important in connection with aspiration or drainage in case of pleuritic effusion, whatever be the character of the latter. Other lines described by authors, but less frequently mentioned perhaps, are the mammillary line, drawn perpendicularly through the nipple on either side; the sternal line, corresponding with either edge of the sternum; the parasternal line, drawn between and parallel with the two preceding; the scapular line, drawn vertically through the inferior angle of the scapula, and the vertebral line, right and left, on each side of the spinal column.

The methods adopted in the physical examination of the chest are, principally, inspection, palpation, percussion, and auscultation. There are other methods of less importance to be described hereafter. The use of the thermometer, or calormetation, as well as the examination of the sputa, are well-known methods of physical examination.

I. Inspection.

Inspection is the act of looking at the patient, and naturally comes first in order. In examining the chest in health the patient should ordinarily be stripped to the waist in a warm and comfortable room, and should stand in the erect position, the heels near each other and on the same line, the arms being dropped loosely by the side. The front of the chest should be inspected first. For this purpose the observer should stand directly in front of the patient and at a convenient distance for inspection.

It is rare to find a perfectly symmetrical chest even in health. The right side may be a little larger than the left, especially in right-handed people with extra development of muscle, as among carpenters and blacksmiths. Not infrequently one shoulder is lower than the other owing to occupation, as is sometimes the case among hod-carriers and tailors; or to previous fracture of the clavicle, or curvature of the spine. Such deviations from the perfect symmetry of the chest may be compatible with perfectly healthy lungs. The apex-beat of the heart may, or may not, be seen, depending a good deal upon the thickness of the chest walls. There is general and even expansion on both sides during inspiration, forced or quiet, and respiratory movements are not usually more noticeable on one side than on the other. Abdominal respiration is more noticeable in men, superior costal respiration in women. The upper part of a woman's chest expands more on inspiration than a man's to allow for child-bearing, the diaphragm in men being a more powerful and important muscle of respiration than in women. On the other hand, abnormal centres of pulsation, the presence of tumors, abnormal bulging or flattening of the chest walls, and exaggerated respiratory movements on one side with diminution of those movements on the other, would indicate disease.

For examination posteriorly the patient may be turned around, but should stand in the same erect position. Sometimes a slight lateral curvature of the spine may be noticed, owing to the greater traction of the stronger muscles of one side drawing it in that direction. The shoulders may not be on the same level, as already stated. The scapulæ should move evenly during respiration, as a rule, but there are exceptions. In choreic children, for instance, nervous and hysterical women, or those who chance to be in a nervous condition from the abuse of alcohol, tobacco, or the like, to say nothing of impostors who have heard lectures on the subject, we may find very uneven movements of the chest walls, and especially the scapulæ, though the organs of respiration be perfectly healthy. The uneven movements due to these causes, however, instead of being uniform, as in disease, usually vary and change from one side to the other. When one side steadily and uniformly expands more than the other to any noticeable extent, it is usually indicative of disease, as we shall see hereafter.

In lateral inspection the patient should place the hands on the head. Abnormal bulging or retraction of the sides would indicate disease. But these are observed better from the front, or posteriorly, than the side.

II. Palpation.

Palpation is the act of feeling, and has reference to the special sense of touch. It is the second step in the regular order of examination. It is usually performed by laying the palms of the hands on the patient, but it is sometimes convenient to palpate with the ear in combination with auscultation.

The palms of the hands, when they are used, should previously be warmed, if necessary, and then laid gently, and lightly, on corresponding parts of the chest walls at the same time. This is usually sufficient, but some prefer to apply the hands alternately, or even to cross them, with or without closing the eyes, so as to make the test in every way, in doubtful cases. It is important that the examiner should stand directly in front of, or behind, the patient, according to circumstances, in order to perform this act with proper care. For palpating in front, the patient should stand erect, as for inspection. But when palpating posteriorly, the patient should cross the arms in front and gently grasp the left shoulder with the right hand and the right shoulder with the left hand, keeping the elbows close in to the body, which should be bent slightly forward. In this position the scapulae are moved out of the way and the tissues on the back rendered more tense.

Vocal Fremitus.—What is the object of palpation? Chiefly to ascertain the presence or absence of vocal fremitus, which is the vibration, thrill, or jarring of the chest walls caused by the sound of the voice. And if present, to know whether it is abnormally increased or diminished.

Fremitus, or jarring of the chest walls, may be produced in various ways and is designated accordingly. If produced by the voice it is called vocal fremitus or voice thrill. From the fact that the chest walls vibrate it is sometimes called pectoral thrill. As vocal fremitus is the kind of fremitus by far the most commonly observed and referred to, it is often called fremitus simply, without being specified as vocal fremitus. It is more or less generally felt over the whole chest, but more marked on the patient's right side, as will be presently described at length.

Other kinds of fremitus are tussive (or tussile) fremitus produced by the cough, and of use perhaps when the voice is much impaired or lost; rhonchal (or rhonchial) fremitus caused by large râles, or gurgles, and more or less localized at that point; friction fremitus, sometimes felt over a pleuritic friction; and splashing fremitus produced by succussion in case of large cavities containing air and fluid, as in pneumohydrothorax. Besides detecting fremitus we may also, by means of palpation, be enabled to accurately count the number of respirations to the minute, should it be impossible to do this by inspection. For this purpose the hand should be lightly applied to the abdomen in men, or the upper part of the chest in women, for reasons already stated. By palpation also we may be enabled to locate the apex-beat of the heart, and ascertain the character, frequency and rhythm of its movements, as well as of the radial pulse. To a very limited extent we may also conjecture the amount of expansion and contraction of the chest walls during respiration, but these are better told by inspection or measurement if necessary. The surface temperature should be noted.

To determine the presence or absence of the vocal fremitus in any given case by palpation, it is necessary, of course, for the patient to make use of the voice.

For this purpose the patient should pronounce somewhat loudly the words "one, two, three," during the act of palpation, and repeat them as often as necessary. Any short simple phrase would answer the purpose, but these words are as good as any others, and besides they have the sanction of time-honored custom. Some, however, prefer the words, "ninety-nine," or "nineteen." The reason for speaking these words, or some simple phrase, is chiefly because they can be repeated over and over again on the same key by any one. This is very important. In ordinary conversation on any subject the key, or pitch, of the voice being constantly changed, the fremitus varies accordingly. The lower the pitch of the voice the more marked, as a rule, will be the resulting fremitus.

It is important to understand that the normal vocal fremitus is more marked in the right than the left subclavicular region. In other words there is normal exaggerated vocal fremitus in the right sub-clavicular region of the healthy chest. How is this very important fact to be explained? Simply because the right primitive bronchial tube being larger than the left, a larger volume of voice is conveyed into the right lung than the left.

The trachea, about four and a half inches in length, extends from the lower part of the larynx on a level with the fifth cervical vertebra to opposite the third dorsal, where it divides into the two primitive bronchi, right and left. The right bronchus, shorter, more horizontal, and larger than the left, enters the right lung opposite the fourth dorsal vertebra, just behind the upper border of the second costal cartilage of the right side. The left bronchus, longer, more oblique, and smaller than the right, passes under the arch of the aorta and enters the left lung opposite the fifth dorsal vertebra, one inch lower than the right. Moreover, the septum between the two is to the left of the median line, so that foreign bodies getting into the trachea naturally drop into the right bronchus or main chan-

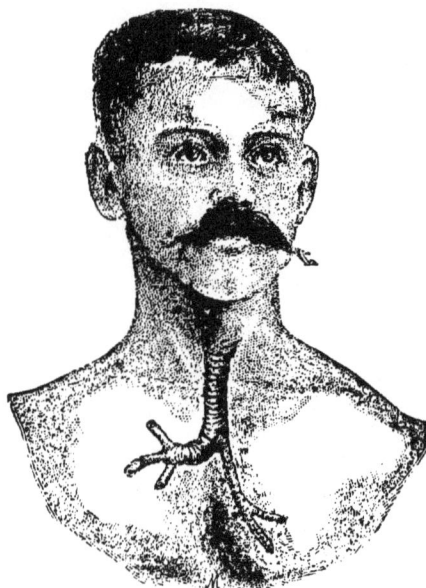

Fig. 4.—Division of the Trachea into the two Primitive Bronchi, showing the Right Bronchus much larger than the Left and given off higher up. (Schematic Diagram.)

nel. It is reasonable to suppose, therefore, that a larger amount or volume of the voice is conveyed into the right lung, especially the upper part, than the left, and hence more vocal fremitus is obtained on the right side. For the same reason more fremitus is felt posteriorly in the inter-scapular region of the right side, as also slightly more in the right sub-scapular region than the left. There being no lung tissue over the superficial area of cardiac dullness in the left mammary region, and over the liver, in front, below the sixth rib, we do not expect to find much fremitus at those points usually, and then only so much as may be extended there by the chest walls. The normal spleen and kidneys do not perceptibly affect the fremitus. Over the scapulæ the fremitus is interrupted more or less by the bone which intervenes.

The amount of vocal fremitus differs in different healthy individuals, as it depends for its production on certain important factors. These may be embraced chiefly under two heads: (1) the character of the voice and (2) the conditions of the chest walls. In the first place, a loud, low-pitched, harsh voice will, other things equal, cause more vocal fremitus than a high-pitched soft voice. For this reason men have more fremitus as a rule than women, and grown people more than children. The bass notes of the large pipes of an organ produce more fremitus, or jarring, than the high notes, the vibrations of the former being more powerful and longer. And in like manner the bass viol produces more fremitus or jarring, than the violin. Secondly, the chest walls. A person with chest walls covered with fat, or extra development of muscle, will, other things equal, have less vocal fremitus than one with thin chest walls, unencumbered with those tissues, thin chest walls requiring less force to be thrown into vibrations. In a man having a loud bass voice and a large chest, with thin chest walls, we should expect to find the vocal fremitus well marked, but more on his right side, for reasons already given. On the other hand, we may not be able to detect any vocal fremitus whatever in a fat, small-chested woman with a high-pitched soft voice. It is of the greatest importance to know that in health the vocal fremitus, usually obtained, is always relatively exaggerated on the patient's right side. The same amount of fremitus on the left side would probably indicate more or less consolidation of lung tissue. In marked deformities of the chest, however, such as those caused by marked spinal curvatures, exceptions to the rule are usually met with in all the methods of physical examinations, though the lungs be perfectly healthy. This is what might be expected, but it is well to bear it in mind. Finally, and what is difficult to explain, the vocal fremitus appears to be very slightly increased when the patient is in the recumbent position, a fact of very little practical importance, as it is not usually noticeable.

III. PERCUSSION.

Percussion is the act of striking the patient one or more blows, to obtain accurate information regarding the underlying parts, and is the third step in the regular order of the physical examination of the chest. It should be performed gently so as not to cause pain to the patient. Although the history of inspection and palpation appears to be nowhere stated, it is well known that AUENBRUGGER was the author of percussion. He was born in Gratz, Styria, in 1722, and practiced medicine in Vienna, where he died in 1809. He discovered the value of percussion while engaged in the study of a case of empyema. He, however, made use of immediate percussion only, and published his views on that subject in 1761. But it

was not until PIORRY of Paris invented the pleximeter
in 1828 that mediate percussion was brought into general use. Piorry also drew attention to the increased
sense of resistance which accompanies the dull sound
elicited by percussing solidified lung tissue, and hence
one of the advantages of palpatory percussion, with
the finger as a pleximeter, over other methods. The
percussion hammer was invented by WINTRICH in 1841,
and when employing this instrument, instead of the

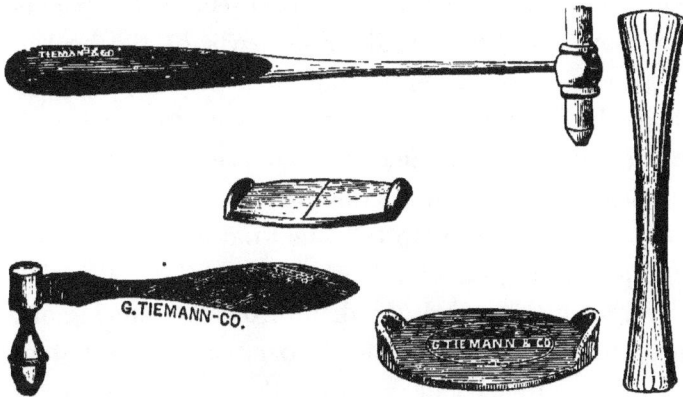

FIG. 5.—Percussion Hammers and Pleximeters.

fingers, it is necessary frequently to use a pleximeter
made of ivory or other material, which may be flat or
otherwise. A solid piece of ivory about two inches
long, a little larger than an ordinary lead pencil, and
shaped somewhat like an hour glass, is useful sometimes in percussing just above and below the clavicles
in cases of marked depression in those regions, and between ribs, in some cases when it would be difficult to
apply the finger, or other pleximeter. To perform palpatory percussion, a finger of the left hand is placed as
a pleximeter firmly against the chest walls and hori-

zontally between ribs, rather than vertically. Then with one or more fingers of the right hand (in right-handed persons) curved so as to bring their tips in contact with the pleximeter, and moving the wrist joint only, three to five, short, sharp blows, are rapidly delivered, which produce a certain sound. The pleximeter should be applied alike on both sides of the chest and the blows should be delivered with equal force. These rules are equally applicable when the percussion hammer and ivory pleximeter are used. The percussion sound differs for different parts of the healthy chest, and is composed of four elements or properties, (1) quality or timbre; (2) pitch; (3) duration or length, and (4) intensity or amount. Of these properties by far the most important is quality.

Quality enables us to tell one kind of a sound from another regardless of all other properties. We thus distinguish the sound of a drum from the blowing of a horn and these two from piano music. It is not by the pitch, intensity or duration, that we recognize the sound of the drum, but by the quality of the sound. And so for all other musical instruments. The drum may be small and give a high-pitched sound. Or it may be large and give a low-pitched sound. Or the pitch may vary with tightening or loosening the head of the same drum. While the pitch may vary infinitely, the quality remains the same.

In percussing over healthy vesicular lung tissue we know that the sound elicited is the normal vesicular, or pulmonary resonance, simply by the quality of the sound produced. The pitch will vary in different healthy people. A person with a large chest will give a lower pitched note perhaps than one with a small chest, and of two chests of the same size, that one will give a higher note the walls of which are more tense or thick than the other. Yet the quality may be the same for all. In percussing over the liver, deltoid muscle, or any solid substance, the quality of the note is dull; over fluids inclosed in thin walls it is flat; and over a cavity with tense walls and containing air it may be tympanitic. If it be a large cavity, the pitch will be low, and if a small one, other things equal, it will be high, relatively, yet it will be tympanitic in both cases. It is the quality of the percussion note, then, that helps us to distinguish the healthy from the pathological condition, and one pathological condition from another.

Pitch as an attribute of chest sounds was first brought forward by Dr. Walter H. Walshe, of London, about 1850. It is not possible for all to distinguish slight differences in pitch, as it is a natural gift, and belongs to those who have, what is called, an ear for music. Even then it requires cultivation and practice. On the other hand, every one can recognize one quality of sound from another. Pitch is often of use in distinguishing different degrees of the same class. Thus a high-pitched tympanitic percussion note gives the idea of a smaller cavity, with tenser walls, than low-pitched tympanicity, although the quality in both cases would be the same. A large area of solidification would yield a higher pitched percussion note than a small one, though the quality in each case would be dull. There are, however, different degrees of dullness, and the more marked the quality of dullness is, the higher will be

the pitch. Where the dullness is very slight the pitch
is of great importance, otherwise it is not.

Ordinary respiration does not affect the percussion
note, since we obtain the average quality of the sound
during percussion. If, however, the patient be directed
to take a deep inspiration and hold the breath, the
note may be a little different then from one made at
the end of a full expiration. In the former case the
resonance is exaggerated, but the pitch varies for different cases, sometimes higher and sometimes lower
than at the end of a full expiration, according to the
tension of the chest walls and the volume of air inhaled.
A chest with little expansive power may be made very
tense on forced inspiration without increasing the volume of air in proportion. Here the pitch would be
higher at the end of inspiration. If volume exceeded
tension it would be lower. It is not surprising, therefore, to find even eminent authors differing on this
rather unimportant point.

Duration, or length, as an element or property of
sound, varies with the pitch. The higher the pitch the
shorter the duration or length, and the lower the pitch
the longer will be the duration.

Intensity, or amount, as already stated, is used in
two senses. In percussion the intensity signifies the
amount in the sense of amplitude or volume. The
lower the pitch, therefore, the longer the duration and
the greater the intensity, or volume, of the percussion
note. Normal percussion resonance therefore is lower
in pitch, longer in duration, and of greater intensity
(volume) than dullness which is higher in pitch, shorter
in duration and of less intensity (volume). A large

cavity giving low-pitched tympanicity emits a more intense percussion note than a small one giving high-pitched tympanicity. All these elements or properties of sound are important, but quality, as already stated, is first, and pitch second, the duration and intensity being secondary to pitch.

Let us now examine the different percussion sounds obtained in percussing different regions of the healthy chest; marked deformities being excepted, as already stated, when speaking of palpation. Great care should be observed in symmetrically arranging the patient so as to be able to compare one side with the other under the same conditions. If we percuss in the left sub-clavicular (infra-clavicular) region a certain sound is obtained termed the normal vesicular resonance, or the normal pulmonary resonance. The quality is vesicular, or pulmonary, due to the presence of normal vesicular, or pulmonary tissue. The pitch is somewhat low, the duration and intensity (volume) being in proportion. In the right sub-clavicular (infra-clavicular) region we obtain a note very slightly duller in quality, somewhat higher in pitch, shorter in duration, and of less intensity (volume). This fact is of the utmost importance, since slight dullness on percussion is one of the earliest signs of incipient phthisis, due to incomplete consolidation. Three reasons are given for this difference. (1) The right pectoral muscles being more developed and thicker in right-handed persons would naturally cause the note to be slightly duller on the right side than the left. Should the patient be left handed the note may be slightly duller on the left side but not always. We then look for some other reason. (2) The right lobe of the liver acting as a solid foundation for the right lung is regarded by some as the cause for a higher percussion note on the right side than the left, the left lung being in relation with the large end of the stomach. The heart being situated chiefly on the left side, however, this theory is disputed and even rejected by some. (3) Lastly the difference is said by some to be due to the difference in the anatomical arrangement of the bronchial tubes. The right bronchial tube being larger, and situated higher up, than the left, gives us broncho-pulmonary, or broncho-vesicular tissue, to deal with in which the bronchial element is more marked than in the same region on the left side. And it is not improbable that bronchial tubes with their muscular coats and more or less connective tissue, occupying space that is taken up in the left side by air cells, would give a slightly duller note. Flint, however, states that they probably raise the pitch by imparting to the note a slightly tympanitic quality of high pitch.

The resonance obtained by immediate percussion on the clavicles is not always indicative of the true condition of the lungs underneath, and hence is not to be depended on. The properties of the note will vary with the length and shape of the bone. For these reasons a slightly duller note is often obtained over the left clavicle than the right, in perfectly healthy chests.

The mammary regions being covered over with more adipose tissue than those above or below in connection with the mammary glands which are also more or less developed, the note in these regions would naturally be less resonant than in those above. Moreover in the left mammary region we have within the line of the left nipple the deep and superficial areas of cardiac dullness, the former extending into the right mammary region about half an inch to the right of the sternum. Along the lower margin of the right mammary region we find the line of deep hepatic dullness.

The right hypochondriac (infra-mammary) region being occupied by the right lobe of the liver we obtain here marked hepatic dullness. It should not be called flatness, however, as that quality is obtained by percussing over fluids contained in thin walls. In the upper part of the left hypochondriac region and at its outer boundary we may obtain some dullness from the left lobe of the liver and anterior border of the spleen, but owing to the larger end of the stomach the percussion note, especially under forcible percussion, is often tympanitic. Over the sternum dullness by means of gentle percussion is usually obtained on account of the bone. But forcible percussion over the trachea and bronchi down to the line of the upper border of the third ribs will give tympanicity or rather bandbox resonance. Below the third ribs the heart and liver cause dullness.

In percussing the posterior regions of the chest the patient should assume the same position as for palpation, so as to get the scapulae as much out of the way as possible and render the tissues tense. In the suprascapular regions forcible percussion elicits pulmonary resonance to a certain extent, but it is muffled by the thick covering of over-lying tissues. This is even more the case in the inter-scapular regions. Over the scapular regions the bone interferes with the resonance. In the sub-scapular (infra-scapular) regions, however, we get pulmonary resonance down to the lower limit of the lungs, which is the tenth rib on both sides. The note, however, is slightly duller and consequently higher pitched in the right sub-scapular region than the left, both on gentle and forcible percussion. The right muscles of the back being usually thicker than the left, and the right lobe of the liver affording a solid substance against which we percuss on the right side, fully account for this difference. The lungs only extend down to the tenth ribs on the vertebral lines (both sides of the spinal column), so that below those ribs there is usually marked dullness. In percussing the lateral regions the patient should place on the head the hand of the side percussed. In the axillary regions of both sides we get loud pulmonary resonance. In the sub-axillary (infra-axillary) region of the right side we come down to hepatic dullness on the axillary line, between the seventh and eighth ribs, and it continues down to the eleventh. In the same region on the left side there is some dullness on percussion over the spleen from the ninth to the eleventh ribs. Forcible percussion in this region, however, may elicit tympanicity from the large end of the stomach.

In children the percussion note is usually more resonant than in adults, owing to their thin chest walls. In women, for the same reason and on account of their superior costal respiration, the resonance is more marked than in men, especially in the upper part of the chest; while in the aged, the chest walls becoming more and more rigid, and the lungs smaller in volume, less resonance is to be expected as a rule.

IV. AUSCULTATION.

Auscultation is the act of listening, and may be done immediately by placing the ear directly against the chest, or mediately by the intervention of a stethoscope. In the former case a thin soft towel, or other similar material, may be used to cover up the chest.

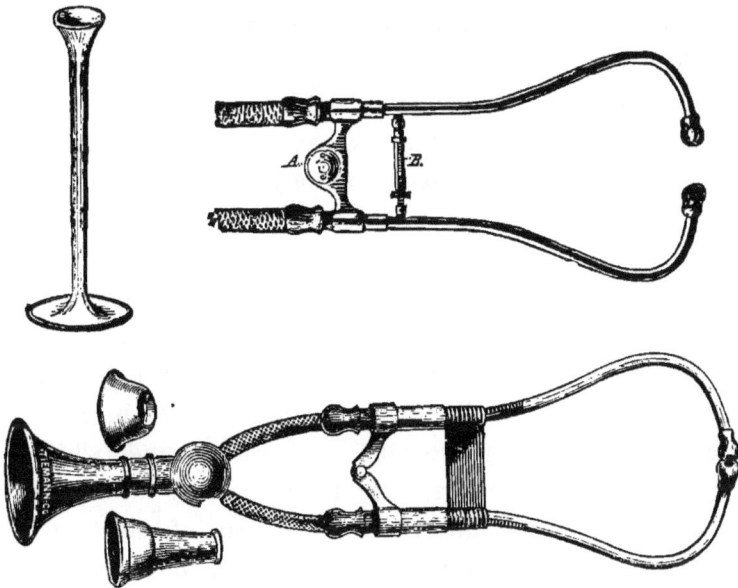

FIG. 6.—Stethoscopes.
A. Ford's silent circular spring.   *B.* Screw for regulating ear-pressure.

The same observations regarding the position of the patient in palpating, or percussing, apply here.

Hippocrates, 460-375 B.C., was the first to make use of auscultation as a procedure in physical diagnosis, but only to the limited extent of hearing the splashing of fluid by succussion, in the case of pneumohydrothorax. It was not until 1816 that Laennec, of the Necker Hospital, in Paris, invented the stethoscope

and first gave to auscultation the value which it now
possesses. In 1840 Dr. Camman, of New York City,
invented the binaural stethoscope. The ear pieces of
this instrument should fit properly, and the spring, or
elastic, may be guarded by a screw so as to regulate
the pressure in the ears, otherwise injury to the ears
must follow sooner or later. The moderate use of the
stethoscope for locating heart murmurs is good, but in
other cases the ear is to be preferred when possible.

In the auscultation of the healthy chest we listen for
the breath sounds and the voice sounds; the respiratory murmur and the pectorophony, or vocal resonance
as heard over the chest. The patient should be directed
to clear the throat if necessary and not to make any
superfluous noise in breathing.

Normal Respiratory Murmur.—If now we place
the ear or stethoscope to the chest while the patient
breathes with a moderate amount of force, we hear the
normal respiratory murmur, or breath sound, which
for reasons already given is more marked in women
than in men, especially in the upper part of the chest.

This normal respiratory murmur is a more or less
composite sound in which the larynx, trachea, bronchial tubes, air vesicles, and perhaps other elements are
concerned. It differs for different regions of the chest
and is designated according to the predominating quality present. In the left sub-clavicular region, for instance, we have the type of the normal vesicular respiratory murmur, the vesicular (rustling, breezy) quality
predominating on account of the presence of the pulmonary vesicles or air-cells. In rhythm the length of
the inspiration of the normal vesicular respiratory

murmur is about four times the length of expiration and continuous with the latter. Inspiration is vesicular (breezy or rustling) in quality, of a certain pitch, which may be regarded as somewhat low, and of a certain intensity, which varies in different healthy chests. The expiration when present is always continuous with inspiration, but is absent, according to Flint, in about one-fourth of the cases. This is especially true with men, particularly on the left side. Expiration is blowing in quality and lower in pitch than inspiration. Guttmann truthfully states that this respiratory murmur may be imitated by properly adjusting the lips (nearly closing them) and drawing in and ex-

FIG. 7.—Normal Vesicular respiratory murmur.

pelling air through them on the proper key with due regard to duration. In fact any respiratory murmur may be imitated in the same way, as in whispering a tune. The reason why expiration is shorter than inspiration, according to Walshe, is because in the latter instance the air current is directed toward, in the former from, the ear of the auscultator. Either inspiration or expiration or both may be wavy in perfectly healthy chests. We find it in the hysterical and nervous. Palpitation of the heart may cause it, especially on the left side. Even when interrupted, cog-wheeled, or jerky, it is not necessarily associated with incipient phthisis.

The normal respiratory murmur heard in the right sub-clavicular region differs somewhat from that heard

in the left. On the right side the bronchus imparts
to it a bronchial element not heard on the left. In
other words we have the normal vesiculo-bronchial
respiratory murmur on the right side, a fact of the
utmost importance, especially when taken with the
other physical signs in this region, all of which very
closely resemble those of incipient phthisis. In this
vesiculo-bronchial murmur the chief characteristics are
that expiration is prolonged, raised in pitch, and somewhat tubular in quality, and the murmur as a whole
is more intense (exaggerated) than it is on the left
side. Inspiration is also slightly higher in pitch and
less vesicular in quality than on the left side.

In the supra-sternal region by placing the stethoscope over the trachea we get the normal tracheal
breathing, and in the inter-scapular region, especially
of the left side, as the left primitive bronchus is more
deeply situated than the right, we obtain the normal
bronchial breathing over the site of the left primitive
bronchus. In this both inspiration and expiration are
tubular in quality. The bronchial respiratory murmur
resembles the tracheal but is less intense. Over the
larynx we have the normal laryngeal breathing. Over
the liver and superficial area of cardiac dullness, where
no lung tissue exists, we hear no respiratory murmur,
unless it may be transmitted along the chest walls.
Over the scapulae the murmur is weak or absent on account of the intervening bone. The normal vesicular
respiratory murmur besides being heard in the left
sub-clavicular region is also heard in both axillary regions and the sub-scapular regions. The respiratory
murmur may also be heard in the supra-clavicular and

supra-scapular regions. In any locality it may become exaggerated, supplementary, or hypervesicular in certain cases where one lung or portion of a lung is doing extra work for a short time. In children before the vesicular element of the lungs has become fully developed and also owing to their thin chest walls, the respiratory murmur is termed puerile. It is less vesicular and more intense than in adult life, but is not necessarily therefore harsh. In old age the murmur becomes less intense and the rhythm changed, inspiration being shortened and expiration prolonged.

Normal Pectorophony (chest voice), vocal resonance, or voice sound over the chest, in health is the sound produced by the patient's voice as heard through the chest walls. It is a distant, diffused, indistinct sound, being more or less buzzing. The ear or stethoscope, preferably the ear, should be applied to the chest and the patient directed to speak out the words one, two, three, repeating them as often as necessary.

Normal pectorophony (chest voice) or vocal resonance, differs in different healthy people according to the character of the voice and the conditions of the chest walls, and for different localities, just exactly like the normal vocal fremitus. What has been already stated about the latter, therefore, equally applies to the former. It is normally exaggerated (slightly increased) in the patient's right sub-clavicular region and elsewhere, like the fremitus.

SUMMARY OF THE PHYSICAL SIGNS OF THE HEALTHY CHEST.

1. *Inspection.*—A perfectly symmetrical chest is rare. The shoulders may not be on the same level, and there may be slight lateral curvature of the spine. The apex-beat of the heart may or may not be visible, according, chiefly, to the thickness of the chest walls. There is equal and general expansion of the chest walls on both sides during inspiration. Superior costal respiration is noticeable in women, abdominal respiration in men. The scapulæ move evenly, except, for instance, in nervous people.

2. *Palpation.*—The normal vocal fremitus varies in different people, depending on the character of the voice and the conditions of the chest walls. It is, however, more perceptible all over the patient's right side than the left and is especially exaggerated in the right sub-clavicular, inter-scapular, and sub-scapular regions, since the right bronchial tube, larger than the left, conducts a larger volume of voice to the right side. The apex-beat of the heart is usually, but not always, felt.

3. *Percussion.*—We obtain normal pulmonary resonance in the left sub-clavicular region, both axillary regions, and the left sub-scapular region. In the right sub-clavicular and sub-scapular regions the note is slightly duller than in those regions of the left side, owing chiefly to the difference in the thickness of the muscles. Over the liver and superficial area of cardiac dullness we obtain marked dullness. Over the spleen in the left sub-axillary region, from the ninth to

the eleventh ribs on the axillary line, there is slight
dullness mingled not infrequently with ventral tympanicity. In the scapular and inter-scapular regions
the resonance is interfered with by the intervening
bone and muscles.

4. *Auscultation.*— In the left sub-clavicular, both
axillary and sub-scapular regions, we hear the normal
vesicular respiratory murmur. In the right sub-clavicular region we hear the normal vesiculo-bronchial
respiratory murmur owing to proximity to the right
primitive bronchus. Over the larynx, trachea and
bronchi we hear the respiratory murmur characteristic
of those organs. Over the liver and superficial area of
cardiac dullness the respiratory murmur may be entirely absent unless transmitted there by the chest
walls. The normal pectorophony (vocal resonance over
the chest) is exaggerated in the right sub-clavicular
region and is also more intense in the right inter-scapular and sub-scapular regions than the left for the same
reason that the fremitus is.

From the foregoing observations we deduce the
following differential summary of the physical signs
as obtained in the right and left sub-clavicular regions in the healthy chest: (1) Inspection is chiefly
negative. (2) Palpation gives exaggerated vocal fremitus on the patient's right side. (3) Slight dullness
on percussion in the patient's right side, the pitch
being slightly raised. (4) Upon auscultation we find
vesiculo-bronchial respiratory murmur and exaggerated pectorophony (vocal resonance) on the patient's
right side. In other words, we have in the right sub-
clavicular region of the healthy chest all the signs of

incomplete consolidation of lung tissue as seen in the early stage of phthisis, except some adventitious sound, such as crepitant, or sub-crepitant râles, or the mucous or intra-pleural click. " Without a practical knowledge of these points of disparity," says Flint, " error in diagnosis can hardly be avoided." In cases of doubt the sputa should be examined by an expert for the tubercle bacillus, though it is not always found at this early stage.

OTHER MEANS EMPLOYED IN PHYSICAL DIAGNOSIS.

Mensuration, or measurement, may be of use in comparing the two sides of the chest in repose, or during inspiration and expiration, or to ascertain the total amount of expansibility of the chest. Even in these cases inspection is usually sufficient. Moreover, a diseased person can sometimes, by practice, expand the chest walls more than a healthy one who has not practiced. It is of use sometimes in ascertaining the size of the chest in proportion to the height. Thus a man 5 feet 6 inches high should measure around the chest, on a level with the sixth costo-sternal articulations, $37\frac{1}{4}$ inches; 5 feet 8 inches high, $38\frac{1}{2}$ inches; 5 feet 10 inches high, $39\frac{1}{2}$ inches; 6 feet high, 41 inches, and so on. Three inches is good expansion of a healthy chest on full inspiration, although the healthiest person may not be able to expand the chest over two inches without practice, while it may be possible with practice to expand it six inches. Various instruments, such as the stethometer, spirometer, and cyrtometer, have been used for mensuration. The ordinary tape-line, however, is good enough for practical purposes.

Auscultatory-Percussion, was invented by Dr. Camman, of New York, about 1840. It combines auscultation by means of the stethoscope with percussion. While the stethoscope is firmly pressed over the heart, for instance, percuss gently at the same time. As soon as the percussion reaches the border of the heart the quality and pitch of the note is changed at once. It is a much more delicate test than the ordinary method of percussion, and is especially useful in diagnosing thoracic aneurism. It may also be used for accurately determining the boundaries of the heart, liver, and other organs of the body. Unless performed correctly and carefully, however, it is misleading. A pleximeter, small enough to be pressed down between the ribs, should be used, otherwise we always obtain the dull percussion resonance of the superficial adipose, or other tissue, instead of that of the underlying organs.

Succussion, invented by Hippocrates, 460–375 B.C., consists in shaking the patient while the ear is at the same time applied to the patient's chest. By this means the splashing sound of fluid characteristic of pneumo-hydrothorax is detected, as will be fully described.

Paracentesis Thoracis, or *Thoracentesis*, in cases of extensive pleuritic effusions, whether of serum or pus, was first brought before the profession in a practical manner by Trousseau, of Paris, about 1835. After his death it fell into disuse, but was revived in 1852 by Dr. Bowditch, of Boston, Massachusetts, who clearly pointed out the indications for the operation, as will be mentioned when speaking of that subject. More recently by means of the hypodermic syringe the exist-

3

ence and character of pleuritic effusions can be established with certainty.

Other less important procedures are autophonia, the method by musical vibrations, and phonometry.

Autophonia was first brought into notice by the late M. Hourmann, who "connected peculiarities in the resonance of the observer's own voice (as he speaks with the ear applied to the chest directly, or by the intervention of the stethoscope) with certain definite conditions of density of the parts beneath."

The method by musical vibrations originated with Drs. Stone and Grabham (London Lancet, vol. i., 1867, p. 114), and consists in "communicating a musical impulse to the air in the bronchial passages by forcibly inspiring through a tube or pitch-pipe containing a free reed. The note emitted is directly conveyed to the parts under observation."

Phonometry is a method described by Baas. It consists in "placing a tuning fork on the surface of the chest or abdomen, and determining by the intensity or feebleness of the tone it gives, the condition of the subjacent organs." The best instrument for this purpose is Blake's tuning fork with a small spring hammer attached. The usual method of percussion, however, is much to be preferred.

Having become thoroughly and practically acquainted with the physical signs obtained by examining the healthy chest, the student is now prepared to consider the physical signs of various diseases.

CHAPTER II.

BRONCHITIS.

Diseases in which the breath and voice sounds are refracted (broken up, diffused) within the lungs as much as, or more than in health, with consequent normal or diminished vocal fremitus, respiratory murmur, and pectorophony.—Obstruction in the bronchial tubes to the convection of sound, with consequent diminution or suppression of the latter.—Adventitious sounds.

BRONCHITIS is the inflammation of the mucous membrane lining the larger or medium-sized bronchial tubes, but when the smaller or ultimate bronchial tubes become the seat of the disease it is called capillary bronchitis. In neither case are the air cells affected unless as a complication. As independent primary diseases they are general or bilateral, local bronchitis being usually caused by and secondary to some other disease, tubercle, aneurism, or neoplastic growth for instance, or surgical injury.

The physical signs of bronchitis, apart from complications, are as follows:

Inspection.—This yields chiefly a negative result. Usually there is no perceptible difference from health in the size and shape of the chest and character of the respiratory movements. Especially is there no difference in the respiratory movements of the two sides, when compared with one another, but both expand alike.

Palpation.—This also gives negative results generally. The vocal fremitus is usually normal, or it may
be slightly diminished if the bronchial tubes are much
obstructed by mucus, so that the voice sound is not
conveyed by them as freely as in health. In rare cases
a large tube, the left primitive bronchus, for instance,
the right being usually too large, or a large branch of
either, becomes stopped up by a plug of viscid mucus,
with diminution or even absence of the vocal fremitus
over the corresponding area. Coughing may remove
the plug and then the fremitus immediately returns.
In case of coarse mucous râles, especially with thin
chest walls, rhonchal fremitus may be present.

Percussion.—This also yields negative results, the
air cells in uncomplicated bronchitis being unaffected
and full of air. Sometimes the resonance may even be
exaggerated owing to the abnormal distention of the
air cells with air. This, however, will be explained
when speaking of emphysema. In old cases of bronchitis where the inflammation has extended so as to
give rise to peri-bronchial thickening or interstitial inflammation, more or less dullness on percussion will
result from those complications.

Auscultation.—The respiratory murmur may be
weakened if the tubes are much obstructed with mucus,
or, if a large tube be plugged up as already described,
the respiratory murmur may be entirely absent for the
corresponding area. Upon coughing and removing the
obstruction the respiratory murmur immediately returns. When heard, however, its quality is the same
as in health as a rule. It may, however, be a little
harsh (rough) if the mucous membrane of the larger

bronchi is much roughened where the tidal air rubs against it instead of gliding smoothly and without much friction, as in health. Expiration will be more or less prolonged in proportion to the amount of obstruction in the tubes, the force of ordinary expiration being weaker than that of inspiration. Pectorophony (vocal resonance over the chest) is usually normal, but may be slightly diminished like the vocal fremitus. Relatively the vocal fremitus and pectorophony are the same as in health in comparing one side with the other.

Râles (rattles) are usually heard in bronchitis and belong to the adventitious or foreign sounds. These never exist until produced by disease. Adventitious sounds are abnormal from beginning and are not modified normal sounds. Their presence, therefore, always indicates some abnormal condition. (See Summary of Adventitious Sounds, Chap. V.) The râles heard in bronchitis may be dry or moist. There are two varieties of dry râles, the sonorous and sibilant. Sonorous râles are loud, low-pitched, dry râles made in the larger bronchial tubes. The sibilant are high-pitched, whistling râles made in the smaller tubes, or in larger tubes if the constriction be sufficient. They are heard in the early or dry stage of bronchitis and asthma, and are due to irregularity in the calibre of the tubes from spasmodic constriction of the muscular coats, or tumefaction of the lining mucous membrane. Sometimes they are also caused, undoubtedly, by the vibration of viscid mucus, and hence are changed with coughing.

Moist bronchial râles are divided into three varieties. Those made in the larger bronchial tubes are large bubbling râles heard both on inspiration and expiration, and are called mucous râles.

Those made in the medium-sized bronchial tubes are called sub-mucous râles. They are also bubbling râles heard both on inspiration and expiration, but are finer and higher pitched than the mucous râles. Lastly we have the sub-crepitant râles made in the finer bronchi.

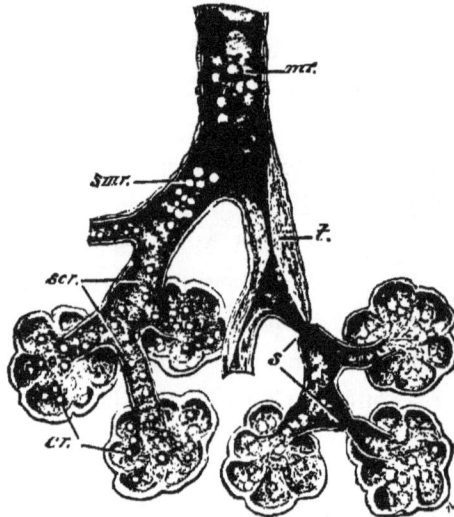

Fig. 8.—Bronchial Râles. *mr*, Mucous râles; *smr*, sub-mucous râles; *scr*, sub-crepitant râles; *cr*, crepitant râles; *t*, tumefaction; *s*, spasmodic stricture of bronchial tube.

These are fine moist râles, heard chiefly on inspiration, though occasionally they are heard on expiration also. All of these râles may be present together, or only one or more varieties. They may be abundant, scant, or even absent at times. In the latter case, coughing may develop them. When present they usually change with or even without coughing, are more or less irregular in size, and are attended with more or less expectoration. In this way they differ from intra-pleural moist râles, which are usually local, do not change with coughing, are uniform and superficial, and attended with little or no expectoration.

The patient should be directed to clear the throat to avoid laryngeal and tracheal râles, the sound of which may be conveyed to the ear through the bronchial tubes and air cells.

Other adventitious sounds are the crepitant râles made in the air cells, intra-pleural râles, mucous and intra-pleural click, gurgles, metallic (amphoric) tinkle, friction sounds, splashing sounds, and certain crackling and crumpling sounds called indeterminate râles, to be .described each in its proper place. (See also Summary of Adventitious Sounds, p. 159, Chap. V.)

Capillary Bronchitis is the inflammation of the mucous membrane lining the smallest or ultimate bronchial tubes, the air cells still remaining unaffected except as a complication. It occurs most frequently in children. It is also seen among the aged and others having enfeebled heart's action, and hence may occur in the course of long-continued diseases of a typhoid character.

The physical signs are, in general, similar to those of ordinary bronchitis.

Inspection, however, may reveal increased frequency of respiratory movements and sometimes more or less cyanosis, especially about the cheeks, ears, tip of the nose, lips, and fingers. (See Atelectasis and Lobular Pneumonia.)

Palpation generally yields negative results, or the fremitus may be diminished or even absent from obstruction of the tubes, as already stated.

Percussion.—The percussion note may be normal, or it may be, and frequently is, exaggerated from temporary over-distention of the air cells. This is due to the fact that forced expiratory efforts empty chiefly the larger bronchi but have very little influence over the air cells. Hence, so far as the air cells are concerned, inspiration is a more forcible act than expiration. The air passes by the mucous obstruction into the air cells more easily than it gets out, and so accumulates in the air cells and distends them. On the other hand, if the patient is very feeble, as is not infrequently the case with sick children or the aged, or those suffering with typhoid fever, the air cells may collapse, giving rise to atelectasis instead of a temporary over-distention or emphysema. In such cases there is not power enough on inspiration to overcome the obstruction in the tubes, and the air in the cells becomes absorbed, the oxygen disappearing first and then the carbonic acid. Such a complication would give some dullness on percussion and is often mistaken for pneumonia.

Auscultation usually reveals the presence of fine sub-crepitant râles in great abundance, over both sides, and especially low down posteriorly, owing to gravitation of the mucus, in many instances, to that part. Other râles may also be present at the same time. The respiratory murmur, if heard at all, would be about normal in quality, with expiration more or less prolonged, owing to the obstacle to the free escape of air. Usually it is so weak, however, owing to its not being freely conveyed by tubes that are obstructed with mucus, that it may be obscured entirely by the râles.

Sometimes the obstruction is so complete that it is suppressed. Pectorophony (vocal resonance over the chest) is usually normal, but may be diminished like the fremitus. If exaggerated, or markedly increased, it would indicate pneumonia.

Differential Diagnosis of bronchitis. This is not usually difficult. Being a bilateral affection with no physical sign of material value, except the detection of various râles more or less over both sides of the chest by auscultation, it is readily distinguished from acute lobar pneumonia which, besides being usually attended with severe constitutional symptoms, is generally unilateral and confined to one lobe, usually the lower lobe of the right lung. Moreover, owing to solidification of lung tissue in pneumonia, we usually obtain marked increase of fremitus on palpation, marked dullness on percussion over the part affected, and on auscultation hear bronchial breathing and bronchophony, as will be fully explained.

In capillary bronchitis the inflammation may extend here and there in spots into the air cells, giving rise to lobular pneumonia, which is also called catarrhal or broncho-pneumonia. It is often difficult to detect minute spots of lobular pneumonia. If in the course of capillary bronchitis, respiration becomes hurried, with exaggerated movements of the alæ nasi and rise in temperature, it is fair to infer that lobular pneumonia has occurred, even if it be to such a small extent that other signs are wanting. But if, in addition, there are increase of fremitus, dullness on percussion, which should be gently performed on children in order to detect it, and bronchial breathing over one or more spots, the diagnosis of pneumonia would be complete. Atelectasis due to capillary bronchitis would also give some dullness on percussion, but it would be symmetrically bilateral and there would be sucking in of the intercostal and supra-clavicular spaces on inspiration, with no extra rise in temperature (see Atelectasis). The presence or absence of chlorides or albumen in the urine in these cases are of little or no value as aids to diagnosis.

ASTHMA.

Asthma is described by authors as a reflex neurosis. It is characterized by recurrent paroxysms of violent dyspnœa, due to spasm of the muscular coats of the bronchial tubes, or tumefaction of their mucous membrane from capillary vaso-motor disturbance, or both. It attacks both sexes at all ages, but men are affected by it twice as often as women, being more exposed to the causes. It requires for its production three factors: (1) a sensitive area of mucous membrane, (2) an abnormally sensitive nerve centre, and (3) an external irritant. Given all three of these factors simultaneously and an attack of asthma is produced, and it cannot be produced by any two without the third.

The sensitive area of mucous membrane may be situated in the naso-pharynx, bronchial tubes, or stomach. The irritant applied to either of these points may be reflected back to the lungs along the pneumogastric nerve, giving rise to naso-pharyngeal asthma, bronchitic asthma or peptic asthma. Cardiac asthma is also described, but this would properly come under the head of bronchitic asthma, owing to the bronchitis resulting from the heart disease. The second factor, the abnormally sensitive nerve centre, may be inherited or acquired by long-continued application of the irritant to the sensitive area of mucous membrane. Not only this, but one sensitive area of mucous membrane may give rise to a second area, so that naso-pharyngeal, or even peptic asthma, for example, may become bronchitic asthma. Of external irritants, which constitute the third factor in this disease, dust of some sort is the most common. But there are a great variety of external irritants, some of them giving different names to the disease, as hay asthma, and the like.

Naso-pharyngeal asthma is very closely allied to what is known as hay fever. In the latter case the sensitive area of mucous membrane in the nares is situated more anteriorly than in asthma, and by reflex action the branches of the facial nerve are involved, giving rise to lachrymation, sneezing, and other symptoms of hay fever. Sometimes hay fever and asthma may occur at the same time. The irritants in cases of peptic asthma are certain kinds of food, and in bronchitic asthma, fog, cold with dampness, and changeable weather, often give rise to an attack.

The physical signs of asthma during an attack are as follows:

Inspection shows labored respiration. Usually the patient sits up and leans forward in order to breathe more freely. The shoulders are elevated and brought forward, the countenance is pale or dusky, and during respiration the expiratory act is seen to be prolonged. During inspiration, which is usually deferred, short, and sometimes jerking, the supra-clavicular, and supra-sternal fossæ, scrobiculus cordis or pit of the stomach, and the intercostal spaces are more or less sucked in and depressed, because the lungs do not become inflated to a degree corresponding to the enlargement of the thoracic cavity. After the paroxysm is over, nothing abnormal is to be observed, unless there be some complication, as vesicular emphysema.

Palpation.—Rhonchal fremitus may be detected owing to the râles (rhonchi) which are present.

Vocal fremitus may be unchanged, or slightly diminished, owing to the emphysematous condition of the lungs rendering them more refractive of sound than in health. It may be also diminished by the obstruction in the bronchial tubes from spasm and mucus, thus rendering the tubes imperfect conveyors of the voice sound.

Percussion.—Exaggerated resonance on percussion is the rule, owing to the emphysematous condition of the lungs.

Auscultation.—Dry râles of various kinds are heard, but chiefly on expiration, which is much prolonged, while inspiration is so short as to scarcely give time enough for these râles to be thoroughly developed. Sometimes they are loud enough to be heard by the patient and the friends in the room. As the paroxysm declines, and increased secretion occurs, various moist râles may be heard, the mucous, sub-mucous, and sub-crepitant. Sometimes moist râles are present with dry râles from the first. The râles are usually heard over both sides of the chest, or they may be heard only on one side, or here and there in spots, but changing about from place to place. The respiratory murmur, if heard, would be little changed in quality, if any, but it is usually obscured by râles, or it may be weakened or entirely suppressed, owing to obstruction in the tubes and refraction in the lungs. The rhythm of respiration is changed, expiration being much prolonged. The reason for this, as well as the emphysematous condition of the lungs, is the obstruction in the bronchial tubes partly from spasm and partly from mucus.

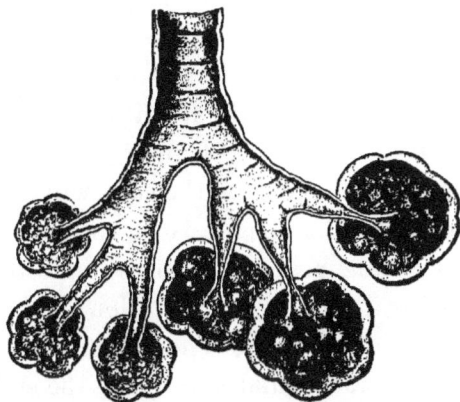

Fig. 9.—Diagram showing Enlargement of the Air-cells during a paroxysm of Asthma owing to spasm of the Bronchial Tubes.

The force of inspiration as affecting the air-cells is greater, as already stated, than the force of expiration. Hence air passes in by the obstruction in the finer tubes more easily than it passes out. Hence its accumulation in the cells behind the obstruction and hence the prolonged effort in the attempt to expel it. It is not the inspiration of fresh air so much as it is the effort to expel impure air that constitutes the chief difficulty in a paroxysm of asthma. Instead of inspiration being to expiration as 4:1 as in health, the rhythm, as Walshe observes, is often reversed and is as 1:4, with expiration sometimes loud and hissing, owing to spasm of the bronchi.

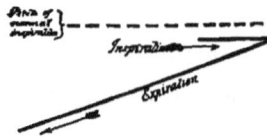

FIG. 10.—Respiration Murmur in Asthma.

After the paroxysm is over, bronchial râles of various kinds, both dry and moist, may be present for a few hours, or even several days.

Pectorophony (vocal resonance) may be normal or somewhat diminished like the fremitus.

Differential Diagnosis.—Proper attention to the physical signs and history of the case will usually enable one to make the diagnosis of asthma with certainty. The suddenness of its onset and departure, the labored breathing with expiration markedly prolonged, the râles so abundant and loud, the absence of purulent expectoration and fever, and the suddenness with which it sometimes yields to remedies, at once distinguish it from bronchitis of any kind. In emphysema the dyspnœa is constant instead of paroxysmal, owing to organic change, besides other physical signs described in connection with that disease (see Emphysema). In some cases of heart disease there are paroxysms of dyspnœa, but they are usually shorter than those of asthma and are unattended with the characteristic prolonged expiration of the latter disease. In spasmodic and other affections of the larynx there is change in the voice, and if wheezing be present, it is located at that point. Moreover, in laryngeal spasm, inspiration and not expiration is prolonged and labored. This may be true of polypus in the trachea. In pulmonary œdema there is dyspnœa, but there is also some dullness on percussion over the seat of the œdema, which is usually situated on both sides at the most dependent portion of the lungs posteriorly. Over the same region also may be heard liquid crepitant râles at the end of inspiration. The dyspnœa, instead of being sudden or paroxysmal, comes on gradually, increasing with the œdema, which may be due to heart disease or associated with dropsy. In hydrothorax, which is a symptom of general dropsy, there is flatness on percussion over the pleural cavities (up to the level of the fluid) instead of extra resonance, and the dyspnœa gradually, but steadily becomes more and more urgent with increase of the dropsy, as in œdema. In pulmonary hypostasis occcuring in old age with enfeebled heart, the dyspnœa is like that due to œdema.

EMPHYSEMA OF THE LUNGS.

Emphysema, or inflation of the lungs, is of two classes: (1) interstitial, where the air escapes through rupture into the interstitial pulmonary tissue; (2) vesicular, where there is excess of air in the pulmonary vesicles, causing them to be abnormally distended.

Interstitial emphysema (inter-lobular, extra-vesicular, extra-alveolar) occurs from rupture of the walls of the air-cells as in the expiratory efforts of violent coughing, in whooping-cough for example, also from parturient efforts, blows, falls, and wounds of the lungs from various causes. Or it may occur in connection with softening and breaking down of tubercle. The

air escapes into the connective tissue of the lungs between the lobules and under the pleuræ. Sometimes
it passes through the posterior mediastinum into the
subcutaneous tissues of the neck, face, and sides, where
it crepitates under the hand during palpation. It is
impossible to make a diagnosis unless the air reaches
the subcutaneous tissue, and in no case is there any
special treatment for it. The escaped air soon becomes
absorbed.

Vesicular (alveolar) emphysema consists of two varieties: (1) vicarious emphysema, and (2) general emphysema.

Vicarious (compensating) emphysema affects one lung
or part of a lung from over work due to crippling of
the other lung or part of lung. In old pleurisy, for
instance, where extensive effusion or adhesions renders
the affected side almost immovable, the other lung becomes vicariously emphysematous from over work. In
lobular pneumonia or collapse of some air-cells, other
air-cells near by may become emphysematous in the
same way. Even in acute lobar pneumonia the unaffected lobes of the same side may become temporarily
or acutely emphysematous to a certain extent. In pulmonary consumption also, crippling of certain areas of
lung tissue may give rise to slight emphysema of
others; not marked, however, as respiration in consumption is usually shallow and rapid, and the volume
of blood much diminished. All vicarious emphysema
is evidently brought about by excessive inspiration.
There is no obstruction to expiration necessarily, and
the patient endeavors to inhale as much air as if both
lungs were intact. Hence inspiration is overdone, with

consequent abnormal distention of the remaining vesicles, and the resulting vicarious emphysema will be
acute or chronic according to the condition that gives
rise to it.

General (substantive) emphysema is also of two varieties: (1) small-lunged and (2) large-lunged emphysema.

Small-lunged emphysema (atrophous, phthisical, senile) is usually found among those well advanced in
years and it may or may not follow the large lunged
variety. The volume of the lungs has become much
reduced owing to atrophy of the intercellular tissue,
though the air-cells have become larger, in many cases,
by coalescence. It is simply a condition mostly belonging to old age.

. General large-lunged (hypertrophous) vesicular emphysema remains for consideration. It is a very serious disease and occurs in men more frequently than in
women, the former being more exposed to the causes;
and in northern, cold, changeable climates more frequently than in warm regions of a uniform temperature, owing, perhaps, to the greater prevalence of bronchitis and asthma in the former. Owing to the length
of time required to produce the disease, it is rarely
seen before thirty years or middle life. Fat people
appear to be more predisposed to it than the thin.
Heredity undoubtedly plays an important part among
predisposing causes. The disease is usually brought
about in one of four ways: (1) by obstructed forcible
expiration, (2) forced inspiration, (3) deformity of the
chest, and lastly (4) by inherited predisposition.

(1) Obstructed forcible expiratory effort is by far the
most frequent cause. For this reason it is found among

4

laborers who are engaged habitually in heavy lifting or straining. In this act the glottis is closed and the abdominal muscles are contracted upon the intestines with great power, which forces the diaphragm up and compresses the air into the upper lobes of the lungs, so that in such cases the upper lobes are chiefly affected, particularly, according to Flint, the left. In the same way players on wind instruments and persons affected with chronic bronchitis, especially the dry bronchitis associated with gout, in which violent fits of coughing habitually occur, may have the disease. The act of coughing simply consists of more or less sudden and violent expiratory efforts with the glottis closed. Children not infrequently have more or less temporary emphysema from whooping-cough.

(2) In some cases the disease is caused by forced inspiration. Here a chronic bronchitis extending to the smaller tubes precedes the emphysema instead of being developed afterward. Inspiration being more forcible than expiration, as far as the air-cells are concerned, the air passes in by the obstruction of viscid mucus more easily than it passes out, so that it accumulates in the air-cells and distends them. In these cases we find the lower portion of the thorax enlarged sometimes, as well as the upper.

(3) According to Freund, in some cases the characteristic deformity (barrel shape) of the chest in large-lunged emphysema occurs first, the lungs afterward becoming distended so as to fill up the vacuum. Such cases, however common in Germany, are very rare in America, if indeed they ever occur. In some cases of deformity of the chest, due to rickets, for instance, or

curvature of the spine from some cause, a local emphysema may occur, one lung or part of a lung becoming distended to fill out one side or portion of the thoracic cavity enlarged by bulging.

(4) Lastly, in certain instances the disease may occur without any mechanical or other known cause, but is simply due to inherited predisposition. This consists in an inherent weakness of the cell walls, perhaps from defective innervation and nutrition, and they yield, in the ordinary avocations of life, without any abnormal force applied.

The dyspnœa and abnormal state of the heart that necessarily accompany marked cases of this disease require brief notice.

There are three factors in the production of the dyspnœa: (1) rigid dilatation of the thorax, (2) loss of capillary area in the lungs, and (3) crippling of the diaphragm. The lung tissue has lost its resiliency and the costal cartilages have become permanently elevated, everted, and hardened, the spaces between them being widened. They have lost their elastic recoil, so that the chest is in a state of rigid dilatation. For this reason air is not expelled properly, and impure air, therefore, accumulates in the lungs causing dyspnœa. In the second place, many capillary blood-vessels in the lungs become obliterated by over stretching. Indeed many of the alveolar septa, on which clusters of these capillaries hang, undergo wasting and perforation, and finally disappear. From loss of capillary area the blood is brought in less quantity in contact with the air in the cells, impure as it is. This second factor is, therefore, a potent one for the production of dyspnœa. More-

over as the blood now has fewer channels left for it in
which to pass through the lungs, the right ventricle of
the heart has more work thrown back on it and hence
becomes enlarged (dilated and hypertrophied). As the
disease progresses, should relative insufficiency of the
tricuspid valves ensue, jugular pulsation and cardiac
dropsy may occur. Lastly, the crippling of the diaphragm is probably the most important factor in the
production of dyspnœa. The volume of the lungs
being increased sometimes enormously, the diaphragm
is permanently pushed down, as is also the heart.
Owing to more or less gastro-intestinal catarrh from
passive hyperæmia due to obstruction to the venous
circulation, there is always more or less dyspepsia with
wind in the stomach. The diaphragm, the most important muscle of respiration a man possesses, is thus put
between two splints. From want of use it becomes
atrophied.

The physical signs of general large-lunged (hypertrophous) emphysema are as follows:

Inspection.—The countenance may be more or less
dusky or cyanotic in proportion to the extent and
progress of the disease. In the later stages jugular
pulsation, due to tricuspid insufficiency, may be observed at the root of the neck on the patient's right
side (see Jugular Pulsation). As obstructed forcible
expiratory effort is the most frequent cause, so the
upper intercostal spaces are usually seen to be widened
owing to the fixed elevation and eversion of the upper
costal cartilages and ribs, like the elevated slats of a
shutter. The upper part of the sternum is prominent,
bulging forward, so as to increase the antero-posterior,

or sterno-vertebral diameter. The shoulders are elevated and brought forward owing to habitual dyspnœa,
and the spinal column is more or less anteriorly curved.
The whole makes up the so-called barrel-shaped chest.
Where the disease has been caused by forced inspiration, the lower portion of the thorax is also enlarged
with widening of the lower intercostal spaces also.

Epigastric pulsation, due to enlarged and lowered
right ventricle, is usually noticeable. The apex-beat
is generally carried downward and outward, but frequently cannot be found, being buried under lung tissue. During forced respiration the thorax moves up
and down more or less like a fixed case, instead of expanding and contracting as in health. During the effort
of deep inspiration the supra-clavicular spaces may be
drawn in, the lungs not expanding any more while
the thoracic walls are raised up. During violent coughing a tumor is sometimes seen to rise up from each of
the flattened supra-clavicular fossæ. This, according
to Niemeyer, is due to the sudden filling up of the sinuses of the jugular veins during coughing, which immediately become empty when the cough ceases. It is
more probably due to lung tissue suddenly distended
upward during violent coughing.

Palpation.—The lungs, owing to distention of the
air-cells, become more positive refractors or diffusors
of sound than in health. Hence the vocal fremitus is
usually diminished, and sometimes even absent. Moreover, owing to bronchitis, which is almost invariably
present in emphysema, either as a cause of the disease
or result of the obstruction to the pulmonary circulation, the bronchial tubes may be more or less obstructed

with mucus, thus rendering them imperfect conveyors of sound. Rigidity of the chest walls and loss of resiliency of lung tissue also lessen the vibrations.

If the bronchitis gives rise to such complications as peri-bronchial thickening, with spots and threads of interstitial induration, the fremitus will be increased over corresponding localities. on account of the more homogenous and better conducting medium for the voice sound (see Solidified Lung Tissue).

Epigastric pulsation is usually felt, and sometimes the apex-beat of the heart displaced down and out. According to some authors it is carried inward toward the epigastrium, but it is difficult to see how it could get there unless the heart be raised up. The cardiac pulsations, and consequently the radial pulse, are often intermittent in this disease, owing perhaps to the extra work thrown on the heart, so that it has to rest every now and then. Dyspepsia, due to gastric venous congestion, and acting reflexly along the pneumogastric nerve, may also cause it.

Percussion.—Resonance is exaggerated on percussion, especially over the upper lobes in front and particularly on the patient's left side. Flint terms this resonance vesiculo-tympanitic. That is to say it is neither purely vesicular nor tympanitic, but a mixture of the two. Biermer, of Zurich, Switzerland, calls it band-box resonance. This is especially the case over the lower posterior and left lateral regions. The pitch of the percussion note in this disease is, according to some authors high, while others maintain that it is low. The truth is the pitch varies in different cases according to the tension of the walls and the volume of air contained. If the walls are tense without much increase in volume of air the pitch will be high. If volume is greater in proportion to tension the pitch will be low. But the quality will be the same in all cases. Variations in pitch, while the quality is the same, may be simply illustrated in a number of ways. The bass

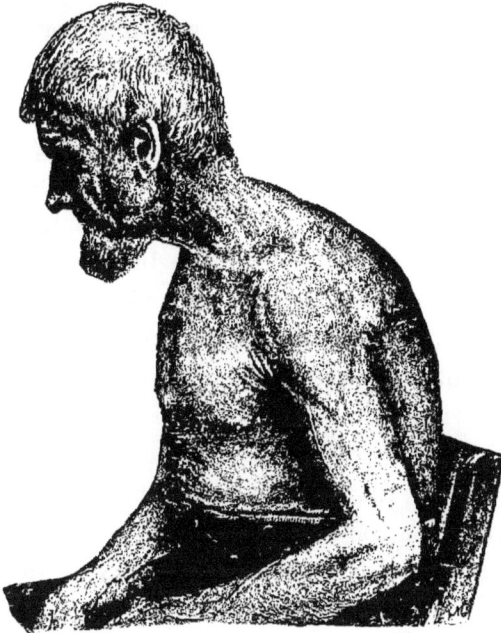

FIG. 11.—General Hypertrophous Vesicular Emphysema.

drum, for instance, yields a lower-pitched tympanicity than the snare drum, yet the quality of tympanicity is the same for both. Drums of the same size with equal tension elicit tympanicity on percussion and the pitch of the note will be the same for all. But if the tension varies the pitch will differ accordingly, greater tension producing higher pitch than if the tension be diminished. The same drum will give a higher or lower pitched note on percussion, according as the head is tightened or loosened. The quality of the note in every case, however, will be tympanitic.

The superficial area of cardiac dullness is diminished or may be absent altogether, only deep dullness on forcible percussion remaining.

In certain cases of atrophous emphysema in advanced life, with hardened costal cartilages and peri-bronchial thickening from long-continued bronchitis, the percussion note in some places may be dull or even wooden in character, especially if the percussion be gentle.

Auscultation.—The lungs being more positive refractors of sound even than in health the respiratory murmur will be weakened, and if the bronchial tubes are, in addition, obstructed by mucus, owing to the bronchitis usually present, the respiratory murmur may be absent. Not unfrequently it is obscured by bronchial râles of various kinds. When the murmur is audible the inspiration is somewhat shorter than in health by being deferred, that is, the first part of inspiration is not heard, being too feeble. It is somewhat lowered in pitch usually and is continuous with expiration, which is often prolonged, not so much from obstruction as from the weakening of the expiratory forces. The diaphragm is permanently depressed, the pulmonary tissue has lost its resiliency and the costal cartilages their elastic recoil. These expiratory forces have become so weakened, therefore, that expiration has to be performed chiefly by contraction of the muscular coats of the bronchial tubes, which are non-striated or involuntary muscular tissue, and have become more or less hypertrophied. Consequently expiration is prolonged, but otherwise it is relatively the same as in health, being lower in pitch than inspiration and blowing in quality. ' Should spasm of the tubes exist due to asthma, with which the disease is frequently associated, the expiration will be about four times longer than inspiration, otherwise it will not be so much prolonged.

Owing to the distended condition of the lungs the heart sounds are usually muffled and feeble. But on account of hypertrophy of the right ventricle the second sound of the heart may sometimes be heard more distinctly (accentuated) over the pulmonary than the aortic inter-space. As dilatation of the ventricle progresses, however, the accentuation becomes less. Should tricuspid insufficiency occur, the corresponding murmur may be heard over the ensiform cartilage (see Tricuspid Regurgitation).

Bronchial râles of various kinds, as already stated, may be present in varying quantity.

Pectorophony (vocal resonance) is diminished, as a rule, for the same reasons that the fremitus is less. When the vocal resonance and fremitus are increased or vary, it is due to some complication, as stated when speaking of palpation.

Differential Diagnosis.—Pneumothorax, or air in the pleural cavity, is the only disease that might be mistaken for emphysema, but even here a careful attention to the physical signs and history of the case renders the diagnosis usually easy. General vesicular emphysema affects both lungs, whereas pneumothorax is nearly always unilateral. Emphysema is developed gradually, pneumothorax comes on suddenly. In the latter disease exaggerated respiratory movements are

observed on the unaffected side, while these movements
on the affected side are diminished or almost entirely
absent, with more or less bulging on that side. In general emphysema there are the barrel-shaped deformity
of the chest, with the thoracic walls moving up and
down as a solid case during respiration. In general
emphysema the heart is displaced downward and usually outward, with epigastric pulsation due to the lowered and enlarged right ventricle. In pneumothorax
the heart is displaced laterally, as a rule, and in a
direction opposite to the pressure. Percussion yields
tympanicity over pneumothorax, whereas the resonance
is only exaggerated (vesiculo-tympanitic, band-box)
and distributed over both sides of the chest in general
emphysema. On auscultation the respiratory murmur is
changed in rhythm in emphysema, and weakened, but
in pneumothorax it is usually absent over the affected
part. Pneumothorax from any cause is always an
acute affection, coming on suddenly and lasting only a
few hours or days. Effusion then takes place giving
rise to pneumo-hydrothorax (or pneumo-pyothorax), to
be detected by the splashing sound heard on succussion.
Emphysema is more slowly developed, is usually
chronic, and for these and the other physical signs
already mentioned, vicarious emphysema, affecting
one side, is readily distinguished from pneumothorax.
Pleurisy, pneumonia, and hydrothorax, although causing dyspnœa, also yield dullness on percussion and
other physical signs altogether different from those of
emphysema. In phthisis there is also dyspnœa, but
the signs of consolidation of lung tissue are entirely
different from those of emphysema.

ATELECTASIS.

Atelectasis (apneumatosis, pulmonary collapse) is a disease characterized by collapse or imperfect dilatation of the pulmonary vesicles, and is the very opposite condition to emphysema.

It is usually situated at the periphery and not the interior of the lungs, but otherwise the site and extent of area differ according to the cause in each particular case. Atelectasis may be congenital or acquired. In the first place the fœtal lungs are in a physiological state of atelectasis, but this disappears as soon as the child is born and breathes freely, all the conditions being favorable. Anything, therefore, that interferes with the respiration of the child at birth may cause more or less congenital atelectasis. Among these causes may be mentioned premature birth, in which case there is not only weakness of the muscles of respiration, but also a want of irritability of the respiratory centre. Accidental plugging of the respiratory tract with mucus, binding the newly born child too tightly, so that the movements of the diaphragm are interfered with, and prolonged and complicated labor, including accidents to the cord, may give rise to congenital atelectasis.

Acquired atelectasis may be due to (1) obstruction, (2) compression, or it may be (3) marasmic.

1. Atelectasis due to obstruction is sometimes the result of capillary bronchitis occurring in weak infants.

The calibre of the tubes is diminished by the swollen mucous membrane and becomes obstructed by the secretion of fluid due to inflammation. Owing to weakness of the inspiratory muscles of such weak infants the obstruction is not overcome by inspiration, and the air already in the cells becomes absorbed, the oxygen first, and then the carbonic acid. Collapse of the cells results. But where the child is strong enough, or among adults, emphysema results instead of atelectasis, as inspiration would then be strong enough to over-come obstruction, but the air could not escape—expiration being weaker than inspiration with regard to the air-cells. In these cases atelectasis is found on both sides over the lower and posterior parts of the lungs, and extending usually in a narrow space up by the sides of the spinal column, disappearing toward the apices.

Other causes of obstruction may be blood clots, fibrinous exudations, bronchial stricture, and pressure on a tube by enlarged lymphatics or other tumors. The atelectasis would then occur in areas correspond-ing to the distribution of the compressed tubes.

2. Atelectasis may be due to compression of the periphery of the lungs, as in pleurisy or pericarditis with effusion, enlargement of the heart, aneurismal or other tumors, hydrothorax, and deformities. Here the site of the atelectasis will, of course, depend on the cause in each case.

3. Finally, we may have what Eichhorst terms maras-mic atelectasis. Whatever diminishes the irritability of the respiratory centre, and weakens the muscles of respiration will contribute to the atelectatic state. Hence we sometimes find it in typhoid or other pro-longed and wasting fevers, paralysis, and brain affec-tions. In such cases the position of the body should not be allowed to remain unchanged during too great

a length of time. Otherwise certain parts of pulmonary tissue, from want of respiratory movement, become
more or less devoid of air, which has become partly or
wholly absorbed, leaving atelectasis. The most dependent portion of the lungs are the most frequently
affected in marasmic atelectasis.

Physical Signs.—In congenital atelectasis, inspection usually shows retraction of the epigastrium and
sinking in of the intercostal spaces on inspiration.
This is owing to the fact that the lungs fail to expand
sufficiently to fill up the thoracic cavity during inspiration, and hence the yielding portions of the chest walls
are sucked in by that act. The breathing is rapid and
shallow, with the interval between inspiration and expiration instead of between expiration and inspiration
as it is in health. Palpation shows no increase of vocal
fremitus as a rule, the collapsed air-cells still acting as
a refractive medium of sound, thus differing from
solidification due to inflammation or compression, when
it conducts sound with corresponding increase of vocal
fremitus. On gentle percussion there is some dullness,
not as much as in solidification. If atelectasis be extensive, slight tympanicity from bronchial tubes may
result. On auscultation the respiratory murmur is
weakened or suppressed instead of being bronchial, as
would be the case in solidification from inflammation
or compression. This is also due to the fact that the
collapsed cells refract sound instead of conducting it.
Hence also, pectorophony (vocal resonance) is also
diminished or weakened. Occasionally, according to
Walshe, there is a little dry rhonchus, probably due to
coincident bronchitis.

In acquired atelectasis the physical signs depend upon the cause. In obstruction and marasmic atelectasis various râles may usually be heard, in addition to the signs already mentioned, owing to the bronchitis present. Obstructions would also be an additional cause for weakened or diminished pectorophony, fremitus, and respiratory murmur. But in compression atelectasis, the physical signs are more like those of solidified lung tissue, in which the part affected becomes a conductor instead of a refractor of sound (see Physical Signs of Lobar Pneumonia). Palpation, accordingly, shows increased vocal fremitus over the compressed lung, dullness on percussion, and bronchial breathing with bronchophony on auscultation. The physical signs of the cause of the compression, as aneurism, pleurisy, and the like, will also be present.

Differential Diagnosis.—Pneumonia is attended with fever, atelectasis is not. Retraction of the epigastrium and intercostal spaces during inspiration is more noticeable in atelectasis than in pneumonia. The bronchial breathing and bronchophony of pneumonia are not observed in any but compression atelectasis, and then the cause of the compression will be apparent. The dullness on percussion in all but compression atelectasis is usually slight and symmetrical. In lobar pneumonia it is unilateral, and in lobular pneumonia a spot of dullness on one side does not necessarily, or usually, have its exact counterpart on the opposite side, as Loomis truly states. According to Grailey Hewitt, extensive deposit of miliary tubercle may be mistaken for atelectasis, but the former is accompanied by fever and emaciation, and perhaps the parents of the child have a tuberculous history. In hemorrhagic infarction the etiology differs, and besides the percussion dullness, and rāles, there is bloody expectoration. Compression atelectasis might be mistaken for one of its causes, pleurisy with effusion. But the effusion occurs at the bottom of the thorax, the compression of lung above. Over the effusion there is a well-marked line of dullness (or flatness) with diminution or absence of the respiratory murmur, vocal fremitus and resonance, and these signs often change with position of the patient. Over the compressed lung the fremitus and resonance would be increased, as the compressed lung would be a better conductor of sound than in health. The breathing, instead of being absent, would be bronchial. To set all doubt at rest, exploratory puncture with the hypodermic syringe may be resorted to. In all cases of compression atelectasis, the cause should be ascertained, if possible.

PULMONARY CONGESTION AND ŒDEMA.

Pulmonary hyperæmia (or congestion) is characterized by excess of blood in the lungs and may be (1) active, (2) passive, or (3) hypostatic.

1. Active hyperæmia (congestion, affluxion, or fluxion) of the lungs, may affect any part of the lungs. It may be caused by direct irritation of lung tissue either from the action of cold or inhalation of irritants, or anything that will cause inflammation. Or it may be due to excessive heart's action from any cause like hypertrophy, emotion, stimulants, or violent effort. It occurs also as a collateral fluxion, some capillaries being over-distended, due to obstruction in others, as seen

in the immediate vicinity of inflammatory foci, in pneumonia for instance. It may also be caused by rarefaction of air in the lungs, as occurs in croup on account of violent efforts at inspiration with the glottis obstructed.

2. Passive hyperæmia is caused by mitral obstruction or regurgitation. In the former the blood is prevented from escaping from the lungs properly, and in the latter it is forced back upon the lungs. In either case the pulmonary capillaries are over-filled. This is usually described as mechanical or obstructive hyperæmia, but as it is also passive there is no necessity for a separate description of it. This form of congestion often leads to brown induration of the lungs (cardiac pneumonia so-called). Enfeebled heart's action, as occurs in typhoid fever, puerperal fever, pyæmia, and in certain centric nervous diseases favors the occurrence of passive hyperæmia, because the left heart having more work to do than the right, fails to empty the pulmonary capillaries as rapidly as they are filled. Hence they become congested.

3. Hypostatic hyperæmia (congestion) is also a passive congestion due to enfeebled heart, but has an additional cause for its production, and that is the retention of the body in one position for a long time, as may occur in typhoid fever, paralysis, and fractures, especially among the aged, requiring them to remain in one position for a long time. Hypostatic congestion is so named because it is a stasis of the under part, or a congestion affecting the most depending parts, and hence is usually found posteriorly in both lower lobes. Hence the necessity for changing the position of such patients occasionally.

The physical signs of pulmonary congestion are similar to those seen in the first stage of lobar pneumonia, before any exudation has taken place.

Inspection.—Dyspnœa will usually be observed, and it will be marked in proportion to the amount of congestion. In certain cases of hypostatic congestion, however, owing to want of irritability in the respiratory centre, as may occur in the course of protracted illness of a typhoid character, dyspnœa may not be present to noticeable degree. In both passive and hypostatic congestion there may be more or less cyanosis. The sitting, rather than recumbent posture, is usually preferred by the patient.

Palpation.—The vocal fremitus is usually normal or even diminished, as the air-cells are in a slightly emphysematous condition, owing to tumefaction of mucous membrane and diminution in calibre of the entrances into the air passages, or termini of the bronchioles. For this reason air enters somewhat more freely than it escapes, as similarly occurs in asthma. The lungs, therefore, being equally as good, or better refractors of sound than in health, the fremitus, as already said, will in some cases be diminished.

Percussion.—As might be expected from the slightly emphysematous condition of the affected part, the percussion resonance may be exaggerated. Very often, however, it is normal. Tympanicity is spoken of by some authors, but it is rare. That quality of percussion sound might be obtained in case of very thin chest walls and marked tension of the pulmonary tissue. Should cracked-pot resonance be obtained it would have no connection with the condition of the lungs,

5

but would be due to concussion of air in the bronchi, as sometimes occurs in children in health and others with thin and yielding chest walls, or to concussion of air in the naso-pharynx.

Auscultation.— The respiratory murmur may be weakened for obvious reasons, or it may be normal, depending on the amount of obstruction at the entrance into the air-cells. Expiration may be slightly prolonged as in asthma. Pectorophony (vocal resonance) would be normal or diminished like the fremitus. The only adventitious sound that would be due to the congestion, would be a very fine sibilant râle heard on expiration, and this may be absent. Any other adventitious sound would indicate complication, such as œdema, or the occurrence of inflammatory exudation, both of which would give rise to the crepitant râle heard at the end of inspiration.

Differential Diagnosis.—This, according to Walshe, is based on the association of sudden dyspnœa and general anxiety with the positive and negative physical signs just enumerated, slightly increased circular measurement of the chest, due to moderate distention of air-cells, exaggerated percussion resonance, and the absence of signs of pleurisy and pneumonia.

PULMONARY ŒDEMA.

Pulmonary œdema, or dropsy of the lungs, occurs when the pulmonary vesicles become the seat of a serous transudation. In some cases the fluid extends into the smaller bronchi, and in others the interstitial tissue may become infiltrated. As a result of the last condition splenization (Loomis) of the lung occurs.

Usually the transuded fluid is colorless, but sometimes it is rose colored from being tinged with blood. Pulmonary œdema is not a primary disease, but is always secondary to, and symptomatic of, some other condition, congestion from some cause as already described, or general dropsy. In the latter case a transudation of the watery parts of the blood takes place, not from the pressure of hyperæmia (congestion), but from the morbidly increased permeability of the blood-vessels.

The location of the œdema will be influenced by the cause in each case. It may be confined to a small spot at any part of one lung, or it may extend over a lobe or entire lung, or even both lungs. Usually, however, it is found low down posteriorly on both sides, as the causes which place it there are most frequent—passive and hypostatic congestion—and also because it is more readily discovered there than in other localities. It may evidently occur in both sexes at all ages.

Inspection, as in congestion, shows more or less dyspnœa, according to the extent of the œdema. The dyspnœa is more urgent than in congestion for the same amount of lung tissue involved. Cyanosis may be observed in some cases. The patient usually prefers the sitting posture.

Palpation.—The fremitus is usually unchanged, as the air-cells still contain some air and may be even weakened as in congestion. According to Walshe, however, in well-marked cases the fremitus may be slightly increased.

Percussion yields more or less dullness as a rule. Before the œdema is well marked the air-cells may be slightly more distended with air than before, as in congestion. For this reason the percussion note may be exaggerated. But as the cells become more filled with fluid and contain less air, the percussion note becomes duller. Tympanicity and cracked-pot resonance are among the possibilities here as elsewhere.

Auscultation.—The respiratory murmur and pectorophony (vocal resonance) are either normal or weakened, as in congestion, or else increased in intensity in well-marked cases. During inspiration, loud, liquid, crepitant râles are heard over the site of the œdema, and this is the chief physical sign. They are bubbling râles made in the air-cells and are not intrapleural. In some cases, sub-crepitant râles are also heard.

Differential Diagnosis.—The chief points of difference between the physical signs of congestion and œdema of the lungs are that in œdema, slight dullness on percussion and loud liquid crepitant râles on auscultation are obtained. In congestion the percussion resonance is exaggerated and there are no râles until œdema or exudation occurs. In the latter case the crepitant râle is much finer and not so liquid as in œdema. From hydrothorax œdema is distinguished by the fact that the line of dullness in hydrothorax usually changes with position of the patient. The dullness in hydrothorax is also much more marked and may even be flat, and crepitant râles will not be heard over the seat of the transudation. In capillary bronchitis there is more or less fever and the sputa are different, being more tenacious and scant than in œdema. Moreover in capillary bronchitis the percussion resonance is exaggerated instead of dull.

CHAPTER III.

Diseases in which the breath and voice sounds are conducted to the chest walls with consequent increased vocal fremitus, respiratory murmur and pectorophony.—Bronchial breathing and bronchophony.—Solidified lung tissue.

PNEUMONIA, or pneumonitis, signifies inflammation of lung tissue, and consists of three varieties according to the pathological condition: (1) lobar, (2) lobular, and (3) inter-lobular, pneumonia. Each of these varieties is also known by other names, as will be seen. Hypostatic pneumonia, so-called, may be either lobar or lobular (see Hypostatic Congestion of the Lungs).

LOBAR PNEUMONIA.

Lobar pneumonia, so named from the fact that this variety of the disease usually affects a whole lobe, or even more, may be primary or secondary. It is commonly an acute disease, although in somewhat rare cases it may become subacute or even chronic. It is characterized by inflammation of the mucous membrane lining the air-cells, and this inflammation in some cases may even extend up into the bronchioles, the reverse process of what occurs in lobular pneumonia. From the character of the exudation it is sometimes called croupous pneumonia, a term first applied by Rokitansky. According to Virchow and others, however, this term should only be applied to those

cases that result from laryngeal croup), and in other
cases, the exudation being fibrinous, it should be called
fibrinous pneumonia.   On the other hand, Hoffmann,
Flint, and some French authors, regard neither as correct, and suggest pneumonic fever as the true definition.
The disease is known among the laity in New England,
and other parts of the United States, as lung fever.
Again, owing to its affecting the parenchyma, or secre-
ting structure of the lung, it is sometimes called parenchymatous pneumonia, although this is true also of
lobular pneumonia.

Acute primary lobar (croupous, croupal, fibrinous,
exudative, parenchymatous) pneumonia, peripneumo-
nia or pneumonic (lung) fever, is said by some to be an
infectious disease or specific fever, of which the lung
affection is only a local manifestation.   By others it is
claimed to be a local inflammation with resulting symptomatic fever.   It is also yet a matter of dispute as to
why it usually affects the lower lobe of the right lung.
Advocates of the infectious theory say that the right
primary bronchus being larger than the left, pneumo-
cocci are drawn into the right lung in greater abundance than into the left, and naturally drift downward
toward the most depending portions.

The order of frequency with which the different lobes
are attacked is as follows: lower lobe of right lung,
lower lobe of left lung, middle lobe of right lung, upper
lobe of right lung, upper lobe of left lung.   Or it may
extend from one lobe to another on the same side, or
it may attack two lobes on different sides.   In the
latter case it is said to be double pneumonia.   In cases
where an upper lobe is primarily attacked, it usually

occurs among the aged or those addicted to intemperance, especially just after or during a debauch, pneumonia potatorum (Huss). Exposure incident to the intoxicated state, added to the debilitated condition caused by hard drinking, may in some measure account for it.

The disease affects both sexes at all ages, but men more frequently than women, owing chiefly to difference in habits, occupation and mode of life. In more than three thousand cases collected by Barry, nearly five times more men than women were affected. The proportion is usually estimated at about three to one. According to Schramm this proportion is reversed in old age. However this may be, it appears that the difference is not so marked at those ages when the sexes live under similar conditions.

In regard to age, according to Grisolle, Wilson Fox and others, although lobular pneumonia, when it does occur, is found chiefly among children and old people, yet these two classes are subject also to lobar pneumonia as well.

Lobar pneumonia is, according to the same authors, very frequent in infancy, especially during the first two years of life, less common between infancy and twenty years of age, quite frequent from twenty to forty, less so from forty to sixty, and very frequent after sixty years of age. According to Loomis, nine-tenths of all deaths after the sixty-fifth year are caused by lobar pneumonia. Lowering vitality from any cause, such as improper and insufficient food, exhaustion from overwork, intemperance, or previous illness, and living in ill-ventilated and damp apartments, predisposes to it. One attack also predisposes to a second or more, although subsequent attacks are generally not so severe as the first.

Cardiac diseases that obstruct the pulmonary circulation favor an attack. The disease is more common in variable climates than in those of uniform temperature, and hence is not met with in the tropical or polar but in the temperate regions, and hence also it is more prevalent at certain seasons of the year. Epidemic influenza appears to exert an influence in the production of pneumonia. Exposure to wet and draughts of cold appears to act as an exciting cause in some cases.

Inhalation of chemical irritants, injuries, and foreign bodies in the air passages, may also act as exciting causes in a small proportion of cases. Tilten found that in 320 cases, only 4.5 per cent. were connected with traumatism.

Secondary lobar pneumonia occurs as an intercurrent affection in the course of some exhausting disease, as chronic malaria, Bright's disease of the kidneys, diabetes melitus, and also in such diseases as measles, scarlet fever, small-pox, erysipelas, typhoid and typhus fever, rheumatism and pyæmia.

Hypostatic pneumonia, when it occurs, succeeds hypostatic congestion, which is a passive congestion taking place in the most dependent portions of the lungs, and hence is frequently bilateral. It may be lobar or lobular (see Congestion of the Lungs). It is due to imperfect cardiac function from valvular lesion, or cardiac enfeeblement from some cause, and hence is seen in the aged and infirm, or in the course of typhoid fever or

other exhausting disease with the body in one position
for too great a length of time. It may also follow excessive loss of blood from injuries, parturition and the
like.

On the other hand, pneumonia rarely affects the lungs
in emphysema, probably on acount of loss of capillary
area with diminished circulation of blood in the lungs
in that disease.

Stages of Lobar Pneumonia.—It is usually divided
into three stages, not counting incubation, which, according to different authors, varies from a few hours
to two or three weeks.

The first stage, or that of congestion, varies usually
from a few hours to twenty-four hours. The second
stage, or that of red hepatization or solidification, lasts
in ordinary cases about four or five days, so that the
crisis, as indicated by the sudden fall in temperature,
occurs from the fifth to the eighth day. The third and
last stage is that of gray hepatization or resolution,
and lasts about eight days, so that by the end of the
second week from the commencement of the disease, the
patient, in ordinarily favorable cases, is rapidly con-
valescent. The duration of the different stages, however, is subject to many variations. The physical signs
usually become apparent within twenty-four to forty-
eight hours from the first symptoms of the attack, but,
according to Wilson Fox, they may be delayed three
or four days, showing that the pneumonia was probably central at first. If it remain central no marked
physical signs may be observed at all, but such cases
must be very rare. The physical signs of the different
stages are usually as follows:

Inspection. First Stage.—Diminution of respiratory movements on the affected side, and exaggerated respiration on the unaffected side. Also marked abdominal breathing if there be no diaphragmatic pleurisy and the diaphragm be left free. Pain in the side is usually present, and, if severe, the respiration will be jerking, the patient favoring the affected side on each inspiration. The respiratory movements are increased in frequency. The apex-beat of the heart, if noticed at all, will be in its normal position unless displaced by some complication. The cheeks are usually red, sometimes only one, and that on the affected side, which is simply a coincidence. The patient may lie on the affected side at first, so as to restrain the movements of that side if the pain is severe, as well as to allow free motion on the unaffected side. Afterward the dorsal decubitus, or half-sitting posture, is usually preferred.

Palpation. First Stage.—The vocal fremitus is normal or even diminished before exudation commences. But soon it becomes exaggerated over the affected lobe, which is rapidly becoming a better conductor and less of a refractor of the voice sound, as the air-cells become obliterated. Complications like thickened pleura, or pleurisy with effusion over the site of the pneumonia, obstruction in the bronchi by pressure from some cause, or from plugging with viscid mucus, or diminishing their calibre by the congestion, would cause the fremitus to be diminished or absent. The skin is usually hot and dry, pulse increased in frequency, 80, 100, 120, and the apex-beat of the heart, if felt, would be in the normal position, unless displaced by complication.

PLATE I.

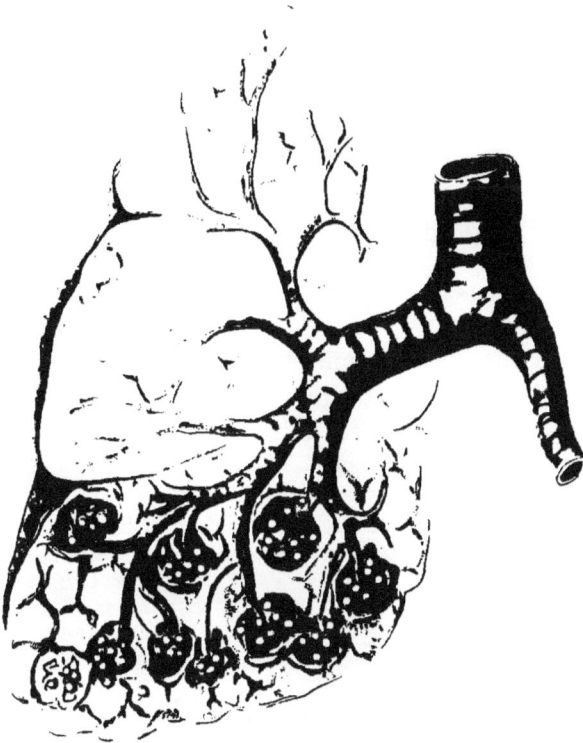

Acute Lobar Pneumonia—Lower Lobe Right Lung. First Stage, Congestion. Crepitant Râle.

BY H. MACDONALD M.D. N.Y.

Percussion. First Stage.—Slight dullness is sometimes observed early in this stage, but sometimes the resonance may be exaggerated or even tympanitic, and occasionally cracked-pot resonance is noticed (see Tympanitic Resonance, and Cracked-pot Resonance). Guttmann and Eichhorst account for tympanicity and cracked-pot resonance in this stage and the third by "relaxation of the pulmonary parenchyma, which is filled with fluid containing air, and sometimes to concussion of air in the bronchi." According to Walshe, exaggerated resonance is more frequent than slight dullness in this stage. If the disease be central or a small area involved there will be no change in the percussion note from health. This diminished vocal fremitus and exaggerated percussion resonance is due to the slightly emphysematous condition of the air-cells as already mentioned, on account of obstruction to the exit of air from the air-cells owing to swelling of mucous membrane in the air passages, which, however, is not sufficient to prevent the entrance of air.

Auscultation. First Stage.—In the earliest period of engorgement the respiratory murmur may be weakened, owing to obstruction due to swelling of the mucous membrane. Soon, however, it begins to lose its vesicular quality and becomes vesiculo-bronchial. Over the healthy portions it may be somewhat exaggerated. Crepitant râles are sometimes, not always, heard over the affected lobe. They are heard at the end of inspiration, very rarely on expiration (Walshe), and are fine crackling râles which sound like rubbing a lock of hair between the thumb and finger over the ear, or like the crackling of fine salt when thrown upon the fire.

There are two theories as to their mode of production. One is that they are produced in the air-cells by the separation of the agglutinated cell-walls at the end of inspiration or by viscid exudation and air in the cells, or stretching of the inter-vesicular tissue; and the other is that they are fine, dry, pleuritic friction râles, and are only present when the pleura is involved. During the latter part of inspiration, when the pleural membranes rub together, the visceral and parietal, instead of their gliding noiselessly on each other as in health, there is produced an abundance of fine crackling friction râles. Both of these views are still held by different authors. It is more probable, however, that the old theory is the correct one, that they are produced in the air-cells, however closely they may be imitated by pleuritic râles. They are uniform in size and are not changed by coughing, nor do they disappear until the second stage, reappearing in the third, and hence termed the râle redux. Pectorophony (vocal resonance over the chest), like the fremitus, may be normal or even diminished at first, but soon it usually becomes exaggerated.

The chief physical sign in the first stage is the crepitant râle, heard at the end of inspiration and over the affected lobe when present, but it is sometimes absent, and very often the disease has passed already into the second stage before the case is seen. Other râles and adventitious sounds may be present, owing to complications

Inspection. Second Stage. — This gives a similar result as in the first stage. The respiratory movements are increased in frequency, being shallow and panting, more or less restrained on the affected side and increased on the other, especially during a full inspiration. Diaphragmatic respiration will be prominent unless restricted by pain.

Palpation. Second Stage.—The vocal fremitus is usually markedly increased over the solidified lobe. This is owing to the fact that the solidified lung tissue affords an unbroken, homogeneous medium for the conduction of waves of sound, instead of refracting them, and hence they reach the chest walls with much greater force, with consequent increase of fremitus, than in health, where the air-cells act as a powerful refracting medium. When, however, a whole lung becomes solidified, with obstructions of the bronchi, so that the voice sound cannot enter the lung, the vocal fremitus may be diminished or even absent. The same is true if the bronchi be obstructed from any cause, or if the thickened pleura, or pleurisy with effusion, exist on the same side.

Percussion. Second Stage.—The note is usually markedly dull in quality over the affected lobe, and consequently the pitch is high, the duration short, and the intensity or amount, in the sense of volume or amplitude, diminished.

The percussion resonance is somewhat exaggerated over the healthy lung tissue, especially over the adjoining border of the next lobe. This is due to a temporary vicarious emphysema which varies in different cases.

The boundaries of dullness and resonance in pneumonia do not change with position of the patient. When the whole lower lobe is affected the line of dull-

ness in front extends upward and outward along the
interlobar fissure. Occasionally the percussion resonance over the solidified lobe may be tympanitic, or
even cracked-pot, according to Guttmann. In the upper
lobes, this Skodaic resonance of tympanicity, obtained
sometimes while percussing over solidified lung tissue,
(or even effusion), is due to proximity to the primitive
bronchi or the trachea, or both. Tympanicity in such
cases, however, would be attended with a sense of resistance on palpatory percussion, which would not be
felt on percussion in the case of pneumothorax, for instance.

Over the lower lobes Skodaic tympanicity is often
due to the fact that the force of the percussion blow
extends to an empty stomach or transverse colon, especially in children, or "relaxed pulmonary parenchyma near the solidified portion." According to Guttmann the cracked-pot resonance, very rarely met with
in this stage, is due to sudden expulsion of air from
the larger bronchi leading to the affected lobe. It might
probably be also due to concussion of air in large
bronchi, or the trachea if the seat of the disease be in
the upper lobes, and is usually noticed in children or
those with thin and yielding chest walls.

Auscultation. Second Stage.—The respiratory murmur as heard over the solidified lung tissue is known
as bronchial breathing. It is simply laryngo-tracheal
breathing conveyed along the bronchial tubes to the
solidified tissue, and by the latter it is raised in pitch
and conducted with considerable force to the chest
walls, unmodified by air-cells. It has, therefore, not
acquired any vesicular, breezy, or rustling quality, but

PLATE II.

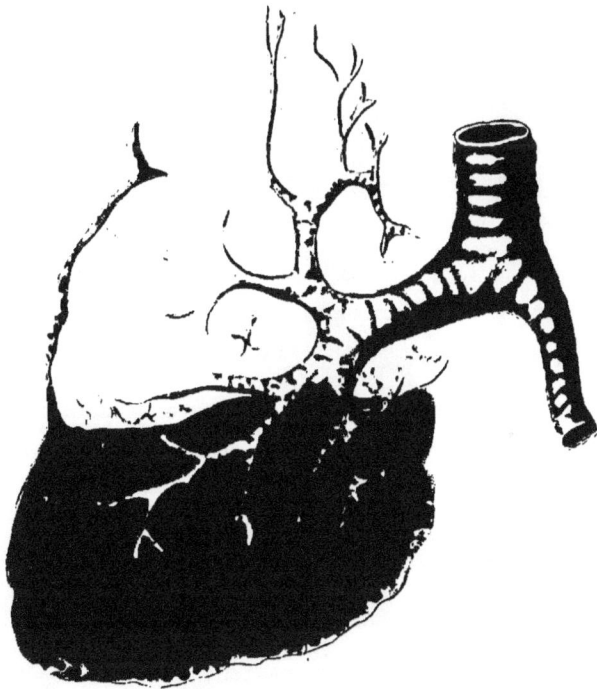

Acute Lobar Pneumonia—Lower Lobe Right Lung. Second Stage, Red Hepatization or Complete Solidification, Vocal Fremitus markedly increased. Bronchial Breathing and Bronchophony.

BY H. MACDONALD M.D. N.Y.

BEATTY & VOYTS, 33, LTH. 163 & '64 WEST ST N.Y.

simply remains intensely tubular usually, and high-pitched, both on inspiration and expiration. The latter is usually higher pitched than inspiration because of the natural conformation of the larynx. Inspiration is unfinished, so that this form of respiratory murmur is said to be divided, and the interval between inspiration and expiration will be marked in proportion to the amount of solidified lung. Inspiration and expiration are, therefore, not continuous, and expiration is as long as inspiration, or even longer. Owing to the shallow, panting respiration in pneumonia, both the inspiratory and expiratory murmurs are, however, rather short.

Bronchial breathing may be represented somewhat by the following diagram:

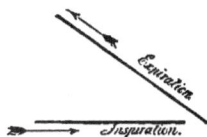

Fig. 12.—Bronchial Breathing.

Bronchial breathing is more intense than the normal respiratory murmur in the sense of concentrated amount. It sounds something like blowing gently across the mouth of the stethoscope. A close imitation of it may be obtained, as Guttmann aptly suggests, by placing a piece of liver, which resembles hepatized lung-tissue, in a tin or other tube, with a thin membrane over each end, and listening through it to tracheal breathing. Although Dr. Walshe and others describe this murmur as being sometimes blowing in quality, it is always more tubular than blowing, and the latter description is misleading. It is always tubular, but more markedly and distinctly tubular sometimes than at others.

Pectorophony (vocal resonance) is greatly increased in intensity in the sense of concentrated amount, not volume; and as the voice usually (not speech) is only heard, it is called bronchophony (bronchial voice). This term was first applied by Laennec, of Paris (about 1820), in contradistinction to his pectoriloquy (chest-speech). The latter is usually low-pitched and refers to the articulate words often heard over cavities. But sometimes, as is well known, we hear the articulate words (bronchiloquy) over consolidation. Should the voice be whispered, as Flint suggested, we should get whispered bronchophony, or whispered bronchiloquy over consolidated lung tissue, as the case might be.

Bronchophony (or bronchiloquy) is necessarily high-pitched, because the voice sound, in case of consolidation, passes from one medium into a denser medium with shorter vibrations and consequently raised pitched. Normal pectorophony (vocal resonance) is a distant, diffused, indistinct, somewhat low-pitched, buzzing sound. Bronchophony (and bronchiloquy) is just the reverse, being near, concentrated, distinct, and high-pitched. When whispered it is tubular in quality, cavernous whisper being often low-pitched, and blowing. In percussion the low note is more intense in the sense of volume, but in bronchophony (and bronchiloquy), which is high-pitched, the intensity is increased in the sense of concentrated amount.

Should thickened pleura or effusion intervene, both bronchial breathing and bronchophony may be weak and distant, and the respiratory murmur may even be entirely inaudible. So may whispered bronchophony. but if the patient speaks loud enough bronchophony will be heard. It will, however, be distant and weak. Aegophony (goat's voice), according to some authors, is heard in this stage, being due to fluid in bronchi that are surrounded by solid tissue. It is quite rare, being more often heard in pleurisy with effusion (see Pleurisy).

The heart sounds may be conducted to the chest walls with markedly increased intensity, especially if the disease be on the left side. No râles or other adventitious sounds are heard unless due to complicating bronchitis or other disease.

Third Stage.—During this stage of gray hepatization or resolution, in favorable cases, there is a gradual return to the normal physical signs. At first they resemble those of the second stage, but soon inspection shows a return of the normal respiratory movements of the chest walls. On palpation the vocal fremitus diminishes over the affected part until it becomes normal, and the marked dullness on percussion gradually yields to normal resonance. Tympanitic and even cracked-pot resonance, as already explained, may, however, occur in this stage. On auscultation, bronchial breathing first becomes vesiculo-bronchial, then normal, and bronchophony yields to exaggerated pectorophony (vocal resonance) which subsequently becomes normal. Subcrepitant and crepitant râles—the latter termed the râle redux, or râle that has come back, having disappeared in the second stage—are usually present until resolution is complete. Abscess and gangrene are rare terminations. In both there are signs of a cavity, with symptoms of general collapse. The former is also at-

6

tended with profuse expectoration of pus, the latter
with fœtid prune-juice expectoration.

Differential Diagnosis.—In cases of central pneumonia with healthy peripheral cell tissue the diagnosis
may be very difficult. Such cases are comparatively
rare, however, and in addition to the history of the
disease the character of the sputa would be of great
service. In œdema and hypostatic congestion loud
liquid crepitant râles are heard, usually on both sides
posteriorly and low down, or wherever the most dependent portions happen to be, and the sputa, though
blood-stained, are abundant and watery instead of the
characteristic rusty colored viscid sputa of pneumonia.
These affections, besides being frequently associated
with some other disease, are also unaccompanied by the
chill, pyrexia and pain of pneumonia, and there is slight
instead of marked dullness, with absence of bronchial
breathing and bronchophony. Lobular pneumonia,
inter-lobular pneumonia, and tubercular consumption,
besides tumors of various kinds, aneurismal, syphilitic,
or carcinomatous, as well as enlargements of the spleen
and liver, may give more or less signs of abnormal
solidification within the thoracic cavity, but absence of
the characteristic sputa of lobar pneumonia, as well as
the general symptoms of that disease, with attention
to the physical signs as already detailed, will usually
lead to a correct diagnosis. The vocal resonance and
fremitus will be diminished or absent, and bronchial
breathing and bronchophony will be wanting usually
over enlarged liver and spleen.

Acute pleurisy with effusion gives different physical
signs from pneumonia, and it is by these alone that a

PLATE III.

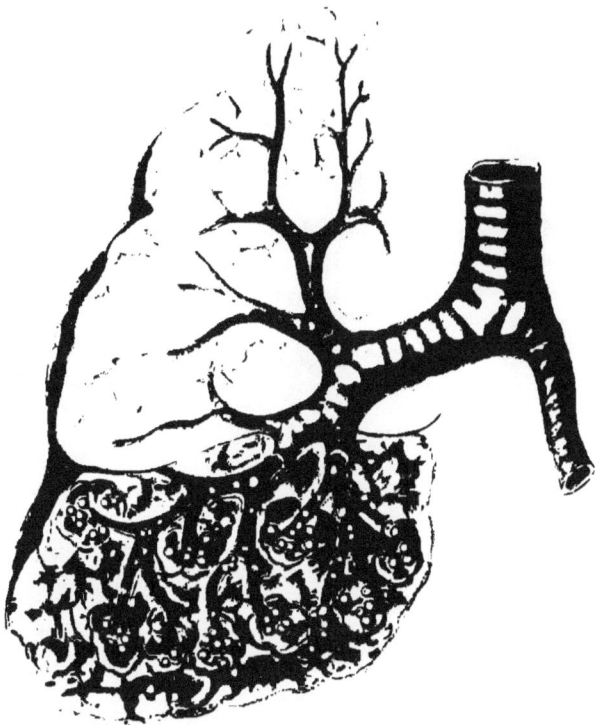

Acute Lobar Pneumonia—Lower Lobe Right Lung. Third Stage, Gray Hepatization or Resolution. Crepitant (Râle Redux) and Subcrepitant Râle.

DRAWN BY H. MACDONALD M.D. N.Y.

positive diagnosis can be made. Otherwise it very closely resembles pneumonia in some respects. In pneumonia the solidified lung tissue acts as a conductor, but pleurisy with effusion (and even thickened pleura) acts as an interrupter of sound, as a diaphragm or partition placed between the examiner and the patient's voice sound. Consequently, while in pneumonia the vocal resonance and fremitus are both markedly increased with bronchial breathing and bronchial voice (bronchophony), in pleurisy with effusion the vocal resonance and fremitus will be usually diminished or even absent, as also the respiratory murmur. The line of dullness or flatness in pleurisy increases downward, and, moreover, often changes with position of the patient, but does not in pneumonia. In cases of doubt the hypodermic syringe as an aspirator would set the matter at rest, but this step is hardly ever necessary.

Hemorrhagic infarction may be attended with crepitant and subcrepitant râles, but it comes on suddenly and usually in connection with heart disease, and sometimes pyæmia. Sudden dyspnœa occurring in these diseases, with a small circumscribed area of dullness surrounded by râles, without notable increase of temperature, the expectoration, if any, being usually dark colored, would indicate infarction. Symptoms of typhoid and typhus fever, meningitis, and delirium tremens, might be the cause of over-looking pneumonia, the real source of the trouble. Hence the necessity of a careful physical examination when such symptoms are observed in children, old people, drunkards, or strangers, whose habits and previous condition are unknown. Secondary and hypostatic pneumonia may

easily escape detection unless physical examination be
carefully resorted to and the patient watched.

LOBULAR PNEUMONIA.

Lobular pneumonia is the inflammation of the lining
membrane of the air-cells of one or more lobules. The
difference between this and lobar pneumonia was first
pointed out by Barthez and Rilliet in 1838, although its
true pathology was not clearly shown until 1844, when
Legendre and Bailly first proved it. Until then it was
frequently confounded with atelectasis. As it is usually secondary to bronchitis and due to the extension
of the latter into the air-cells here and there in spots,
it is also called Broncho-Pneumonia, or Catarrhal Pneumonia. From the fact that it affects infants more frequently than it does adults, it is also called by some In-
fantile Pneumonia. This does not imply, however, that
lobar pneumonia does not also attack infants, for it is
more common even among them, according to many
authorities, than lobular pneumonia. The latter disease also attacks the aged, so that it is met with during
the extremes of life, and among others with enfeebled
or crippled heart from some cause. It may occur, however, at any age, and with either sex. It is not infrequently met with in connection with capillary bronchitis during the course of such diseases as measles,
scarlet fever, diphtheria, small-pox, and typhoid fever,
and it is sometimes secondary to pyæmia. Hypostatic
pneumonia occurring in the course of exhausting diseases, as already remarked when speaking of lobar
pneumonia, is often of the lobular variety, with a tendency, however, to spread. From what has been said,

lobular pneumonia may occur on one side or it may be scattered about over both lungs.

The physical signs are often obscure from the nature of the case. They are usually bilateral, but not necessarily symmetrical and to the same degree.

Inspection usually shows in children panting and shallow respiration, the inspiration being short and expiration somewhat prolonged. There are exaggerated movements of the alæ nasi. The upper portion of the chest heaves, the lower portion and abdominal walls being sometimes drawn in during inspiration, but the supra-clavicular and intercostal spaces are not drawn in so much as in atelectasis. Cyanosis to a variable degree may be observed in cases where the disease extends over a considerable area.

Palpation.—Increased vocal fremitus may be detected if the consolidation is sufficiently extensive, but otherwise it will be unchanged.

Percussion elicits dullness in proportion to the amount of consolidation, and the dullness will be over the seat of the disease, usually posteriorly over the lower lobes. It is not infrequently bilateral, but the dullness is not necessarily obtained over symmetrical areas, as usually occurs in atelectasis. Sometimes exaggerated resonance due to emphysema of cells near by, obscures the dullness, and even tympanicity on percussion, as already explained, is sometimes obtained. The percussion should be gentle, and the finger is the best pleximeter in these cases to detect the feeling of resistance afforded·by solidified lung tissue.

Auscultation.—In addition to the bronchial râles due to the co-existing bronchitis, there may be heard bronchial breathing and bronchophony if there be sufficient amount of consolidation.

Differential Diagnosis.—The disease, if confined to one or a few small foci, may entirely escape detection. If, however, in the course of capillary bronchitis there is a sudden rise of temperature and increased dyspnœa, it is fair to infer that lobular pneumonia to some extent has occurred, even if the usual physical signs are wanting. If the disease is extensive, it may be confounded with lobar pneumonia or atelectasis. From the former it is distinguished by the absence of the characteristic physical signs of that disease, extending over a whole lobe on one side, as well as by the history of the case and course of the disease. In atelectasis, due to capillary bronchitis, the fremitus is not increased, except in compression, and the dullness is not so pronounced as in lobular pneumonia. If a spot of dullness on one side does not correspond to one on the other side, according to Loomis, it is in favor of lobular pneumonia. Instead of bronchial breathing and bronchophony in atelectasis, due to capillary bronchitis, the respiratory murmur is weakened or inaudible, and vocal resonance, if changed at all, is diminished. Subcrepitant râles are not heard as a rule over collapsed lung in atelectasis, but nearly always in lobular pneumonia. In atelectasis the signs are often temporary and may disappear with change of position or deep inspiration, the cells becoming immediately inflated with air again. In pneumonia also the. temperature rises, in atelectasis it does not.

INTERLOBULAR PNEUMONIA.

Interlobular or interstitial pneumonia is inflammation of the interstitial connective tissue of the lungs, and is a chronic disease from the first. Hence it is also termed chronic pneumonia. Owing to the shrinkage that follows usually, it was termed by Corrigan cirrhosis of the lungs. Other names are knife-grinder's or potter's lung, sclerosis, schirrus, and fibroid degeneration of the lungs. According to Niemeyer, it is one of the most frequent diseases. In most cases, however, it has been confounded with fibroid phthisis (p. 107), the tubercle bacillus being now found in nearly all cases of what was formerly supposed to be and described as chronic pneumonia.

Interlobular pneumonia occurs rarely, and is then secondary to some other disease, notably chronic mechanical bronchitis from inhalation of irritating particles, acute lobar and lobular (broncho-) pneumonia, and sometimes pleurisy. Although it may attack both sexes at all ages, it is more commonly found among men at middle life or past, naturally, from difference in exposure to the cause and length of time usually required for its development.

Occurring among coal miners, it is called by Zenker anthrakosis pulmonum. In these cases the affected portion of the lungs becomes black from inhalation of coal dust. From inhalation of iron dust, as among knife-grinders, it is called siderosis pulmonum. In other cases where the dust inhaled arises from the handling of cotton, grain, or tobacco, for instance, or in such occupations as brush-making, carpet or silk weaving, stone cutting and the like, the disease is generically termed by Zenker pneumonokoniosis.

In such cases both lungs may be affected, but more frequently it occurs only on the right side, owing to the fact that the right primitive bronchial tube being larger than the left, more dust is inhaled into the right lung than the left. The disease may also result from extension of chronic pleuritic inflammation or from a prolonged attack of lobar or lobular pneumonia, neoplastic growths, and surgical injuries. In these cases the interlobular induration would be situated according to the locality of the original disease or injury. Finally, according to Walshe, syphilis and alcoholism, if not direct causes, appear to favor the development of the disease.

An acute suppurative interlobular (interstitial) pneumonia sometimes, but rarely, occurs. It cannot be distinguished during life from pulmonary abscess. It may occur in spots or it may burrow along the peribronchial tissue, and is then called pneumonia dissecans (Eichhorst). It occurs in exhausted persons, and may result from severe cases of lobar or lobular pneumonia, empyema, or injuries.

Interlobular pneumonia, like fibroid phthisis, almost invariably leads to more or less dilatation of bronchial tubes—bronchiectasis. These bronchiectatic cavities may not be distinguished even anatomically from those occurring in chronic phthisis, but in the latter case the tubercle bacillus is characteristic.

The physical signs of this disease very closely resem-

ble those of fibroid phthisis (p. 131), to which the student is referred.

PULMONARY GANGRENE AND ABSCESS.

Pulmonary Gangrene, or mortification and putrefaction of lung tissue, may be circumscribed or diffuse,
affecting a lobule, a lobe, or an entire lung. It occurs
more frequently perhaps in the course of lobar pneumonia than any other disease, and for that reason is
found more frequently among men than women between twenty and forty, and on the right side, especially the lower lobe of the right lung. It may, however
be found in either sex at any age. Its seat is on the
surface rather than the interior of the lung. Lowered
vitality from any cause, such as improper or insufficient
food, alcoholism, and bad hygienic surroundings, predispose to it. It is therefore usually found among the
laboring and destitute classes. Besides being met with
in the course of lobar pneumonia, as already stated, it
may result as an extension from putrid bronchiectatic
cavities, or septic bodies or food entering the bronchi
and undergoing putrefaction, as in the case of the late
Emperor Frederic of Prussia. Sometimes it is caused
by septic embolism, as may occur in diphtheria, hepatic
abscess, epidemic dysentery, and pyæmia, and sometimes it is the direct result of surgical injuries.

Inasmuch as gangrene is preceded by inflammation,
the physical signs are those of solidification at first, as
already described, and of broken-down tissue resulting
in a cavity afterward (see Cavities, Phthisis, Third
Stage). As soon as gangrene occurs, there are symptoms of collapse, but a positive diagnosis cannot be

made until communication between broken-down gan-
grenous tissue and a bronchial tube is established so
that the sputa can be examined. The fœtid odor and
prune-juice color of the sputa are unmistakable. In
case of a small area of centric gangrene, however, the
disease may entirely be overlooked.

Differential Diagnosis.—The peculiar fœtid and
discolored expectoration of pulmonary gangrene, containing, as it does, decomposed bronchial plugs and
shreds of pulmonary tissue, usually enables one to distinguish gangrene of the lungs from putrid bronchitis,
empyema that has ulcerated into bronchial tubes, abscess, or phthisical cavities.

The etiology of pulmonary abscess is similar to that
of pulmonary gangrene. In pneumonia it is rare, occurring about once in fifty cases according to Huss, but
more frequently in pneumonia of the upper than lower
lobes, and in this respect it differs from gangrene. The
physical signs, like those of gangrene, are the signs of
consolidation until communication between the abscess
and bronchial tubes is established, when there will be
profuse expectoration of pus, followed by signs of a
cavity (which see). There are also symptoms of collapse, as in gangrene, but not so marked.

CANCER OF THE LUNGS.

Cancer of the Lungs may be primary or secondary—
usually the latter. In the former case the cause is the
same as that of primary cancer occurring in other organs, and is unknown. Heredity is probably the most
important factor in its production, although seventy-
five per cent. of all the deaths among the Schneebergen

Cobalt miners is, according to Eichhorst, the result of primary pulmonary cancer. It develops among them at about the age of forty, and after they have been working in the mines for about twenty years. It is thought to be due to the irritation produced by the inhalation of arsenic in the cobalt ore. For this reason, and because the right bronchial tube is larger than the left, and situated higher up, it affects the upper lobe of the right lung most frequently, and occurs in men oftener than in women, and after several years of exposure to the cause, as in the case of anthrakosis pulmonum and the like (p. 87). It infiltrates the pulmonary tissue, and not infrequently cancerous enlargement of mediastinal glands co-exist.

Secondary cancer of the lungs follows cancer in some other part. According to Walshe, other authors, and my own observation, it is secondary in the lungs, especially where the testicles have been the seat of the primary affection. It usually affects both lungs, unequally of course, and is nodular, the nodules varying in size from a pin's head to a child's head. In many cases there are evidences of cancer in other organs, various glands being enlarged. Encephaloid (medullary) cancer is the most common form of the disease in the lungs, although scirrhus, mixed, and other varieties are also found. Although any form of cancer of the lungs affects men most commonly from twenty to forty years of age, it may be found in both sexes at all ages.

There are other neoplastic growths that are found in connection with the lungs, but cancer is by far the most important. They are fibromata, lipomata, enchondromata, osteomata, melanotic tumors, dermoid cysts, myxomata and hæmatomata. They are rarely distinguished during life, as the symptoms they produce are véry vague.

The physical signs of pulmonary cancer are as follows:

Inspection.—Retraction of the chest walls over the affected parts is usually noticed, with diminution of respiratory movement on that side, especially in case of primary infiltrated cancer of the lungs.

In the secondary, nodular form, bulging of the chest walls with displacement of the heart may be observed, if there be a large tumor. The cancer may also appear externally on the chest walls. The sputa are reddish-brown or currant-jelly, usually, and contain cancer cells.

Palpation.—The vocal fremitus, owing to consolidation from infiltration or pressure, is usually increased. But occasionally, especially in the secondary or nodular form, a large bronchus may be pressed upon by the tumor, and the voice may be so obstructed as not to be conveyed by the bronchial tube to the affected part. In such a case the fremitus would be diminished or absent over the corresponding area. The same result might be obtained by complete displacement of lung tissue by a large tumor.

Percussion.—There is dullness, slight or marked, according to the extent of infiltration or size of the nodules. It is usually irregular in outline and does not change with position of the patient, and sometimes extends across the median line on account of co-existing enlargement of mediastinal glands. In case of breaking down of cancerous tissue or tubercle, with which it is sometimes associated, signs of resulting cavity would be present (see Cavity).

Auscultation.—Owing to consolidation of lung tissue, due to infiltration or pressure, bronchial breathing and bronchophony are usually heard. If a bronchus be obstructed by pressure from a large nodule, however, the respiratory murmur and whispered voice would be weakened or even suppressed and absent for a corresponding area to which the tube is distributed. The same result would also be obtained in case of displacement of lung tissue by a tumor. Should consolidation be incomplete, vesiculo-bronchial breathing and exaggerated vocal resonance only would be heard over the affected part. In case a cavity is formed by breaking down of tissue or dilatation of a tube, the respiratory murmur over it would be blowing in quality, and there would be other signs of a cavity (see Phthisis, Third Stage—Cavities).

Until softening has occurred, râles are not commonly heard in cancer of the lungs.

Differential Diagnosis.—Primary infiltrated cancer may be mistaken for other diseases that cause retraction of the chest, as chronic phthisis, pleurisy with retraction after absorption of the fluid, and syphilitic infiltration of the lungs. Either form of cancer may be mistaken for thoracic aneurism, especially as cancerous enlargement of mediastinal glands not infrequently co-exists. The secondary or nodular form of cancer may even be mistaken for pleurisy with effusion, if the tumor be large enough, indeed they frequently exist together.

Fibroid and catarrhal phthisis usually begin in the upper lobes, and, like primary infiltrated cancer, may affect only the right lung. But in cancer

the disease runs a more rapid course and dullness on
percussion may, as Walshe truly states, extend across
the median line on account of co-existing cancerous
mediastinal growths. The sputa in cancer are characteristic, being like currant jelly and containing cancer
cells instead of tubercle bacilli. The cancerous cachexia
is soon established and cancer may also appear elsewhere. In either disease the heart may be displaced,
but it is usually much more so in phthisis, especially of
the fibroid variety. Pain and dyspnœa occur earlier
in cancer and are also usually more prominent. On the
other hand, râles are heard in the earliest stages of
phthisis, especially of the catarrhal form, whereas in
cancer they are not usually heard before softening of
cancerous tissue occurs. By that time the cachexia
and other signs of cancer, already mentioned, are so
well marked that the diagnosis is comparatively easy.
As the disease progresses the characteristic cancerous
sputa and absence of high temperature in cancer would
aid in a correct diagnosis.

In chronic pleurisy with retraction there is cough,
but with little expectoration usually. The vocal fremitus, respiratory murmur, and whispered voice in pleurisy are diminished or absent, and ordinary vocal resonance, or pectorophony, is also diminished. There are
no inward pressure signs on the trachea and œsophagus
in pleurisy. The dullness in pleurisy, much more regular in outline than in cancer, is usually found at the
base of the lungs, posteriorly, and does not extend
across the median line; and the disease is generally
slow in its progress. The reverse is usually true in
cancer, the physical signs of which, as a rule, are found

in the front and upper part of the thorax, rather than posteriorly and low down. The history of the case, the character of the sputa, the cachexia and the greater severity of the symptoms in cancer would also aid in a correct diagnosis. In pleuritic effusion, besides differences already mentioned, the physical signs often change with position of the patient, which probably never occurs in case of cancer. Should the cancer give rise to pleurisy with effusion, the latter, according to Bowditch, is always bloody. The character of the effusion in doubtful cases is easily told by means of the hypodermic syringe.

In syphilitic infiltration there would be the history of syphilis, other syphilitic lesions would probably be present, and improvement would rapidly follow anti-syphilitic treatment.

Thoracic aneurism usually occurs in men at middle life or past, or in either sex in old age. Cancerous enlargement of mediastinal glands, the lungs becoming invaded afterward, occurs mostly in women between twenty and thirty. But cancer of the lungs with co-existing mediastinal tumors often appears in men also about middle life. Both aneurism and cancer may cause dulness on percussion across the median line and inward pressure signs, and be attended by cardiac displacements and murmurs, but in cancer, besides the cancerous cachexia and sputa, there are often distention of superficial thoracic veins, with œdema of the chest, arms, and face, which never occurs from thoracic aneurism. Moreover, in all cases of cancer of the lungs evidences of cancer in other organs may sometimes be obtained on careful examination.

HEMORRHAGIC INFARCTION OF THE LUNGS.

Hemorrhagic infarction of the lungs is circumscribed pulmonary apoplexy due to embolism. (1.) Most commonly there is plugging of a branch of the pulmonary artery by an embolus, followed by transudation (diapedesis) of blood into the parts formerly supplied by the obstructed twig. (2.) More rarely it is produced by rupture of a branch of the pulmonary artery, due to intense hyperæmia around the obstruction, and fatty degeneration of the vessel. The blood in these cases, according to Rindfleisch, finding its way into the air-cells and bronchi of the part, rapidly coagulates and the process ceases. (3.) Finally, in other cases no known cause for the infarction can be found, all trace of an embolus having disappeared. After complete obstruction of a twig of the pulmonary artery by an embolus, the parts beyond the plug, including air-cells, bronchioles, and interstitial tissue, become the seat of hemorrhagic infiltration, how? According to Cohnheim, this is accomplished by the arterial blood being forced backward, or regurgitated, from the capillaries of the pulmonary veins into the excommunicated capillaries of the obstructed twig of the pulmonary artery. The latter do not rupture usually, but allow a diapedesis (transudation) of blood through their walls, which have become abnormally permeable, owing to the disturbance of their integrity by the embolism. Besides this, they probably allow leakage of arterial blood more readily than venous blood to which they had been accustomed. It was formerly thought that the infarction in these cases was always due to rupture of capillaries, simply from intense hyperæmia, as might occur in mitral disease, without taking the embolus into account. But this did not explain why the infarction was limited to such abruptly and well-defined areas, sometimes a single lobule. The true explanation in these cases is that, owing to mitral obstruction or regurgitation, there follows sooner or later dilatation of the right ventricle. At first it is dilated hypertrophy, but in time dilatation becomes prominent, and not being properly compensated by hypertrophy, the blood-current becomes sluggish. In this condition, thrombi, or firm clots of venous blood, may form in the right heart, especially in the musculi pectinati, of the right auricular appendix, or between the columnæ carneæ of the right ventricle, near the apex. From these thrombi, emboli becoming detached, cause infarctions in the lungs by their direct transmission along the pulmonary artery.

Emboli, besides originating in the right side of the heart, from dilatation, may enter the systemic venous circulation from any part of the body, and, passing through the right heart, produce infarction in the lungs. In fractures, or other severe injuries, of the skull, affecting the diplöe, otitis giving rise to inflammation of the petrous portion of the temporal bone, and thrombosis of the cerebral sinuses from any cause, emboli may enter through the superior vena cava into the right side of the heart. In like manner emboli may become detached from thrombi formed in any of the peripheral veins, as sometimes occurs in typhoid or other fevers; also in cases of thrombosis of uterine or ovarian veins,

7

and even the iliac or renal veins from pressure of large uterine fibroid and other tumors.

When the emboli are septic, metastatic abscess usually occurs, instead of infarction, or rapidly follows it.

Emboli originating in the portal circulation, as sometimes occurs in hæmorrhoids or dysentery, are likely to produce infarctions in the liver. Those coming from the lungs, or left side of the heart, affect the brain, spleen, or kidneys.

Hemorrhagic infarction may only affect a single lobule, or there may be several infarctions simultaneously in one or both lungs. From the fact that the branches of the pulmonary artery become smaller as we approach the surface of the lung, infarctions are found more commonly at the periphery than in the interior of the lungs. For the same reason also, central infarctions are more extensive than peripheral. Naturally the infarcted areas are somewhat pyramidal in shape, with their bases toward the periphery.

The physical signs of infarction are as follows:

Inspection.—There is dyspnœa, which upon inquiry will be found to have come on suddenly and unexpectedly. The dyspnœa will be in proportion to the amount of infarction. Where the latter is small, it may be entirely overlooked. But when extensive, the respiratory movements will be increased in frequency, and the breathing may be labored. The sputa are somewhat characteristic, being brownish-red, and darker colored than those of lobar pneumonia.

Palpation.—The vocal fremitus is usually increased, unless the infarction be central or complicated by pleurisy.

Percussion.—There will be dullness on percussion unless the infarction is central or very small, in which cases there may be little or no dullness.

Auscultation.—In case of extensive peripheral infarction, bronchial breathing and bronchophony may be heard over the seat of the part affected. Crepitant and subcrepitant râles are also usually heard. In some cases there may be no physical signs, owing to the small extent of infarcted area, its central location, or existing complications, like pleurisy, for instance. The heart should always be examined.

Differential Diagnosis.—By means of sudden dyspnœa, localized spots of dullness, if any exist, crepitant and subcrepitant râles, and especially the slight rise in temperature with the characteristic brownish-red sputa, which last for a much longer period than in pneumonia, the co-existence of heart disease, especially old mitral obstruction or regurgitation, and the etiology in general, it is usually possible to make a diagnosis. The sputa of cancer of the lungs resemble those of infarction, but the former contain cancer cells, and there would be the cancerous cachexia, and other signs already mentioned. According to Loomis, the sputa of hydatid disease of the lungs may also resemble those of infarction, but in the former case the discovery of hooklets would be decisive.

PULMONARY APOPLEXY.

Pulmonary apoplexy, or escape of blood into the pulmonary tissue, may be circumscribed or diffuse. In the former case, as we have already seen, when due to embolism, it is called infarction. But it may also occur locally from causes which lead to rupture of pulmonary capillaries, either by over-distending them or weakening their walls, or both, without embolism. These causes are fully considered in speaking of hyperæmia and hæmoptysis (which see). In diffuse apoplexy, the pulmonary tissues become more or less destroyed by the extravasated blood, which has escaped from rupture of one or more large vessels. Sometimes the hemorrhage is confined within the lung substance, at other times, from rupture of,the pleura, the blood may be discharged into the pleural sac.

Diffuse pulmonary apoplexy may be due to surgical injury, or rupture of large vessels due to previous atheromatous degeneration, or in the course of gangrene, cancer, or thoracic aneurism. It occurs more frequently among men than women, for the obvious reason that they are more exposed to the cause. Besides profuse hæmoptysis and symptoms of collapse, bubbling râles of various kinds would be heard over the chest before the blood coagulated. If the patient lived, then after coagulation of blood there would be signs of more or less consolidation. After absorption of the clot and recovery has taken place, the signs would again become normal, unless a cavity, cicatrized tissue, or some other abnormal condition remained. Generally, however, the patient dies before physical examination can be made or any treatment be adopted.

HÆMOPTYSIS.

Hæmoptysis is the expectoration of blood, pure, or mixed with other matters, but always in quantity sufficient to be seen with the naked eye. In order to constitute true hæmoptysis, the blood must come from the larynx, trachea, bronchi, or pulmonary tissue, or, according to Walshe, from any part of the respiratory tracts below the epiglottis. It occurs more frequently among men than women, and from fifteen to thirty-five years of age. It is rare in children and old people.

In general terms, wounds or other injuries, ulcerative processes, over-distention of capillaries from any cause, and weakness of their walls, owing to perverted nutrition, are more or less causative of hæmoptysis, as well as hemorrhage from other organs. In addition to these there are certain local causes to be considered.

Laryngeal or tracheal hemorrhage is not a very frequent source of hæmoptysis, nor is it usually copious. At most the sputa are tinged or streaked with blood. It is due sometimes to severe catarrhal hyperæmia (congestion), but is more frequently caused by ulcerative processes of some kind, such as syphilitic, cancerous, or tuberculous, and sometimes it is due to the presence of aneurism.

Bronchial hemorrhage is the most frequent source of hæmoptysis. Besides ulcerative causes, especially in connection with mediastinal tumors and thoracic aneurism and traumatism, the capillaries may become over-distended and rupture from intense hyperæmia, as in severe bronchitis, mitral obstruction and regurgitation, or excessive action of the heart from hypertrophy or stimulants. Rupture of capillaries from weakness of their walls is seen in the hemorrhagic diathesis (hæmophilia), which appears to be hereditary, scurvy, scrofula, and rickets, and in chronic interstitial nephritis (cirrhotic kidney), with hypertrophy of the

left ventricle and brittleness of the arterioles throughout the body. It may also occur, for the same reason,
in the course of typhoid fever, malarial fevers, and the
exanthemata which sometimes assume a hemorrhagic
character.

Profuse hæmoptysis from bronchial hemorrhage may
also occur in tuberculous patients before there are any
physical signs of tubercle. After the disease has become established, it is usually easy to account for the
hemorrhage, but, according to Walshe, " the very early
hæmoptyses of tubercle remain anatomically unex-
plained." It is, however, due to fatty degeneration or
tubercular ulceration of the blood-vessels.

Vicarious bronchial hemorrhage with hæmoptysis
sometimes takes the place of the menses, but even here
the women are probably phthisical.

Lastly, pulmonary hemorrhage as a source of hæmoptysis is next in frequency to bronchial. They are produced by similar causes and often exist together. In
addition to causes already mentioned, it occurs to a
slight extent in acute lobar pneumonia, and hence the
characteristic rusty colored sputa in that disease. It
may be due to rarefaction of air during violent inspiratory efforts with the glottis obstructed, as in croup.
In fact, intense congestion from any cause may produce it (see Pulmonary Congestion).

It may occur also in the course of hydatid disease,
cancer, gangrene, abscess, and pulmonary consumption
in any stage. After cavities are formed in the latter
disease, large vessels ramifying on their walls, or extending through them, first become aneurismal, and
then may rupture, giving rise to profuse hæmoptysis

which may speedily prove fatal. According to Eichhorst, pulmonary hemorrhage may also be of nervous
origin, and occurs among the insane, in chorea, epilepsy,
hypochondriasis, and in various cerebral and spinal
diseases. It occurs also in pulmonary infarction as
well as diffuse apoplexy of the lungs from any cause,
as previously mentioned (see Infarction and Diffuse
Apoplexy of the Lungs).

The physical signs have reference only to those cases
where an appreciable amount of blood is contained in
the respiratory tracts. They differ during the flow of
blood and after coagulation has taken place.

Inspection during the flow of blood. Besides seeing
the blood expectorated, there may be more or less
dyspnœa, increased frequency of respiration, and pallor
of the surface, if the hæmoptysis is profuse.

Palpation and *Percussion* at this stage usually
give negative results. Rhonchal fremitus, however,
may be present.

Auscultation.—Moist bubbling râles of various sizes
may be heard in the different bronchial tubes. Even
crepitant râles are sometimes heard.

The vocal resonance is generally unchanged. But
after coagulation has taken place, all the physical signs
will usually be those of consolidated lung tissue, and
in proportion to the amount of coagulated blood and
the area involved.

Differential Diagnosis.—It is necessary to distinguish hæmoptysis from bleeding from the mouth and
pharynx, epistaxis, and hæmatemesis.

Careful examination of the mouth and pharynx will
readily exclude them as sources of the bleeding. The

same is true for bleeding from the nose. Even in cases occurring during the night when some of the blood is swallowed, or gets into the pharynx during sleep, there are apt to be evidences of nose bleeding. Moreover the blood hawked up in these cases is dark and mingled with nasal secretions and is unattended by any previous symptoms or cough. In hæmatemesis (vomiting of blood) the blood is vomited up, and is usually dark and clotted, and mingled with food and acid. In cancer of the stomach, however, it resembles coffee grounds from being partly digested, and in ulcer of the stomach the vomiting of fresh, liquid blood, may be profuse. But in all cases hæmatemesis is usually preceded by symptoms referable to the epigastric region, such as severe pain after eating, and nausea. In hæmoptysis the blood coughed up is usually bright red, frothy, pure or mixed with sputa, is alkaline, and hæmoptysis is usually preceded by symptoms referable to the chest, such as a sense of constriction, pain, and a warm tickling sensation behind the sternum. In gastric hemorrhage also, the stools may be black (melæna) from the presence in them of dark blood. Occasionally when gastric hemorrhage is profuse, the blood is alkaline and bright red, as in some cases of ulcer of the stomach, and, on the other hand, in hæmoptysis it may sometimes be dark and clotted, but the symptoms referable to the stomach and chest in each case respectively, will usually lead to a correct diagnosis.

The seat of the hemorrhage in hæmoptysis depends on the particular cause in each case. As bronchial and pulmonary hemorrhage are often due to the same cause and may exist together, it would be very difficult, if not impossible, in all cases to distinguish between the two. Bronchial hemorrhage, however, is much more frequent than pulmonary.

ECHINOCOCCI, OR HYDATIDS OF THE LUNGS.

Echinococci, or hydatids of the lungs, as well as of other organs of the body, occur usually where dogs are plentiful, as in Iceland, Australia and other countries. No locality can be said to be free from the disease where there are dogs and open water for drinking purposes, since it is through infected water that the disease is most frequently contracted. Eggs or scolices of the tænia echinococcus may also be inhaled, or carried to the mouth by the fingers, and so enter the body.

When taken into the stomach, as in drinking infected water, they are carried by the portal circulation into the liver, where they usually locate first, especially the right lobe. From thence they may migrate into the lungs, pericardium, pleural cavities, peritoneum, stomach, and intestines. They generally enter the lower lobe of the right lung from the liver, directly through the diaphragm, and hence the frequency with which they are found in that locality. But instead of this, they may also find their way, by the hepatic veins, into the inferior vena cava, and so through the right heart into any part of one or both lungs. When inhaled they are more apt to attack the upper lobe of the right lung, as I saw with Drs. Gœtz and Riverdin, in the case of a lad eighteen years of age, in the County Hospital in Geneva, Switzerland, during a visit there in the month of August, 1888. This case will be referred to again when speaking of pleurisy (see Pleurisy with Effusion—Diagnosis, p. 147).

The physical signs differ according as lung tissue is compressed or one or more cysts are superficial and extensive. In the latter case, besides dyspnœa, hæmoptysis and emaciation as the case progresses, there may be bulging of the chest walls with displacement of the heart and liver. The dullness may extend across the median line, with diminution or even absence of the respiratory murmur and vocal fremitus over the seat of dullness.

In other cases, compression of the lung may give increased vocal fremitus, bronchial breathing and bronchophony at some point. Should the contents of the cyst be expectorated, signs of a cavity may result.

Differential Diagnosis.—Physical signs of pleurisy with effusion at the upper instead of the lower part of the thorax should always be regarded with suspicion. But a positive diagnosis in any case can only be made when a cyst ruptures and scolices or hooklets are found in the sputa. At the same time if a hydatid cyst is suspected and it is sufficiently near the surface, some of the fluid aspirated by means of the hypodermic syringe may be found to contain scolices and hooklets.

PHTHISIS.

Phthisis is tuberculosis, and as now accepted implies pulmonary tuberculosis, a disease caused by a specific virus, the germ of which is Koch's tubercle bacillus and infectious under conditions to be mentioned. It is usually a chronic disease, the forms of which may be reduced to two according to the tissues chiefly involved: (1) catarrhal and (2) fibroid.

Catarrhal phthisis (caseous phthisis, tubercular lobular pneumonia) is the most common form observed, and it usually appears first at the top of the lungs, but not always. Beginning in the bronchioles, it extends into the air cells and along the air passages. From caseous necrosis and softening, cavities result. This form occurs usually between twenty and thirty, and in women somewhat earlier than in men (p. 113).

Fibroid phthisis also usually begins in the bronchioles, but extends along the lymphatics to the pulmonary interstitial connective tissue, producing shrinkage and bronchiectasis as with interlobular pneumonia (p. 87). It occurs more frequently in men and from thirty to forty, and is usually more chronic in its course than the catarrhal form (p. 131).

Phthisis is said to be acute when its progress is rapid. In such cases it may appear to begin as a lobar pneumonia, but its nature soon becomes apparent (phthisis florida, galloping consumption, tuberculous or scrofulous pneumonia). When asociated with extensive hepatization, the acute form is termed by Williams, of London, acute tuberculo-pneumonic phthisis. When the larynx is primarily the seat of tubercle, it is called laryngeal phthisis, or better, tuberculous laryngitis, although primary tuberculous laryngitis is not a common affection. When hemorrhage is an early and prominent symptom it is sometimes called hemorrhagic phthisis.

Acute miliary (general, disseminated) tuberculosis affects various organs and tissues simultaneously. There is usually previous tubercular lesion from which bacilli in large numbers are absorbed into the blood through

lymphatics or unoccluded veins. It usually runs a
rapid course, and somewhat resembles typhoid fever,
the chest signs present, if any, being those of bronchitis with marked dyspnœa. The temperature usually
runs high, but may fall below normal. The disease is
rapidly fatal as a rule, and often not to be detected
during life. On post-mortem examination, not only
the lungs, but the pleuræ and other organs may be
found to be studded with miliary tubercles.

The following brief summary of conclusions is based
upon the transactions of the Congress for the Study of
Tuberculosis, held in Paris during the month of July,
1888.

1. True tubercle generally (not always) contains the
bacillus that was first supposed to exist by Bouchard,
of Paris, in 1880, but afterward discovered by Koch,
of Berlin, in 1882. All other substances—products of
inflammation in connection with other conditions than
tubercle (caseous for example), and occurring in non-
tuberculous subjects—do not contain Koch's tubercle
bacillus. That is the true difference between tubercle
and such other products. When true tubercle does
not contain the bacillus, it is because the latter has
been liberated and escaped, owing to breaking down of
tissue by pathological process, or else the tubercle yet
contains only germs of undeveloped bacilli. True tuberculous virus, therefore, though it may not contain
Koch's tubercle bacillus, is, nevertheless, capable of re-
producing itself by inoculation, under the usual conditions, by reason of bacillary germs. Any other substance than tubercle that contains Koch's tubercle
bacillus, or its germs (caseous products, for example,

which have become infected by the bacillus, or its germs, as may occur in tuberculous subjects), can also produce true tuberculosis by inoculation. Other substances, however, apart from tuberculous subjects, and not containing Koch's tubercle bacillus or its germs, cannot directly produce true tuberculosis.

Koch's tubercle bacillus thrives best at a temperature of 98° F. to 100° F., and is capable of proliferation between 86° F. and 104° F., although it may live far beyond these extremes. It remains virulent in running water at a temperature of 60° F., for six weeks, and in stagnant water, of the same temperature, for eighteen weeks; in dried sputa for a month. It attacks the lungs by preference, although no tissue is exempt from it.

2. Koch's tubercle bacillus is the cause, not the result, of tuberculosis, inherited or acquired, however this fact may yet be doubted or disputed by some. In case of inheritance, not only may there be narrow chest, feeble health, or other predisposition, but the disease may be actually and directly transmitted to the ovule by spermatozoa infected with bacillary germs.

3. Tuberculosis is a contagious disease under certain conditions, especially those that tend to produce lowered vitality, such as overwork, insufficient food, vitiated atmosphere, and prolonged ill health from any cause, and is transmissible even between man and animals by the digestive tract, as was first proved by the experiments of Chauveau, in 1868, through uncooked tuberculous food, especially cow's milk and beef. Also by the respiratory tract, through infected air, as was proved by Villemin in 1869, and Tappeiner in 1876–77. Hence the necessity for strict sanitary measures as a

means of prophylaxis. True tuberculous virus is also transmissible by inoculation, by subcutaneous injection, or through the integument stripped of its epidermis, as was proved by Villemin in 1869; and also in the following ways: By other mucous membranes than those already mentioned, as the conjunctival and genito-urinary; by dermatoses which destroy the epidermis; through the sudoriparous and sebaceous glands; and, finally, through wounded and absorbing surfaces.

4. In the early diagnosis of doubtful or suspected cases of pulmonary tuberculosis in man, Koch's tubercle bacillus should be sought for in the sputa by an expert, as the general practitioner has neither the time, experience, nor the proper instruments for it. Even then it is not always possible to find the bacillus at first, because it may not yet have been liberated, or only its germs may exist. To verify it, two or three lower animals (guinea-pigs, or rabbits, for instance) may be put under the necessary devitalizing conditions, inoculated with the suspected sputa, and killed at varying intervals after a month or more. If it is the disease in question, some tubercles, containing Koch's tubercle bacillus, will usually be found in various organs.

5. No practical means for destroying Koch's tubercle bacillus in patients has yet been discovered, although various methods have been employed.*

6. The management of tuberculosis in man at the present time should therefore be directed toward two

* As we go to press Koch is said to have discovered the cure for tuberculosis.

objects: (1) Prophylaxis in all its relations to the disease, and (2) treatment by medicines only as symptoms
arise that require it. (Report of R. C. M. Page, M.D.,
Delegate from the New York Academy of Medicine to
the Congress held in Paris, July, 1888, for the Study of
Tuberculosis in Man and Animals. New York Medical Record, Oct. 13th, 1888.)

From what has already been said it is seen that tuberculosis is a disease, the germ of which is the bacillus
discovered by Koch in 1882, and that it manifests
itself in various ways, most frequently as pulmonary
tuberculosis or phthisis. But instead of this, or with
it, sometimes it manifests itself as acute miliary tuberculosis, or tubercular peritonitis, meningitis, or pleuritis, and the like.

Pleurisy is thought by some to be productive of
phthisis. While it may be a cause of interlobular (interstitial, chronic) pneumonia from extension of pleuritic inflammation, it cannot be directly productive of
phthisis unless the pleurisy be of tuberculous origin,
as may sometimes happen. In by far the greater
number of these cases, however, the disease begins in
the lungs and extends to the pleura, the affection of
the latter being only symptomatic of the former which
has been overlooked or not even suspected. Non-
tuberculous pleurisy can only give rise to phthisis indirectly by lowering vitality and limiting respiratory
movements so as to render the patient more susceptible
to tuberculous infection. In like manner chronic pneumonia and bronchitis may act as predisposing causes,
the latter also favoring the lodgment of bacilli by
roughening the bronchial mucous membrane.

The disease attacks both sexes at all ages, but is seen most frequently among those between twenty and thirty. As already stated, anything that tends to lower vitality predisposes to it. In addition to what has already been said under this head, may be mentioned insufficient light, damp localities, previous disease, mental anxiety, rapid child-bearing, lactation, venereal excess, menstrual disorders, and the like. Occupations necessitating inhalation of irritating particles would not of themselves produce fibroid phthisis, although they act powerfully as predisposing causes. According to Flint, those who already have general emphysema, or cardiac disease, are not so subject to pulmonary consumption of any kind.

A much-vexed question has been the relation between scrofula, so called, and tubercle. According to Flint, those who suffer from scrofulous affections of the cervical lymphatic glands are not liable to pulmonary tuberculosis. "Assuming," says Flint, "the scrofulous and tuberculous cachexia to be the same, as is claimed by some, it seems to be exhausted by deposit in the glands of the neck, and is not likely to occur afterward." The truth is, there is no such thing as scrofula, but it is tuberculosis in which the bacilli are confined to the lymphatic system, where they become destroyed or modified. On the other hand, they thrive in the lungs and proliferate rapidly.

In phthisis the disease usually begins at the apices, but as to whether the apex of one lung is more frequently affected than the other is not a settled matter. As in most cases, however, the disease is started by inhalation of air containing bacilli, the apex of the

PLATE IV.

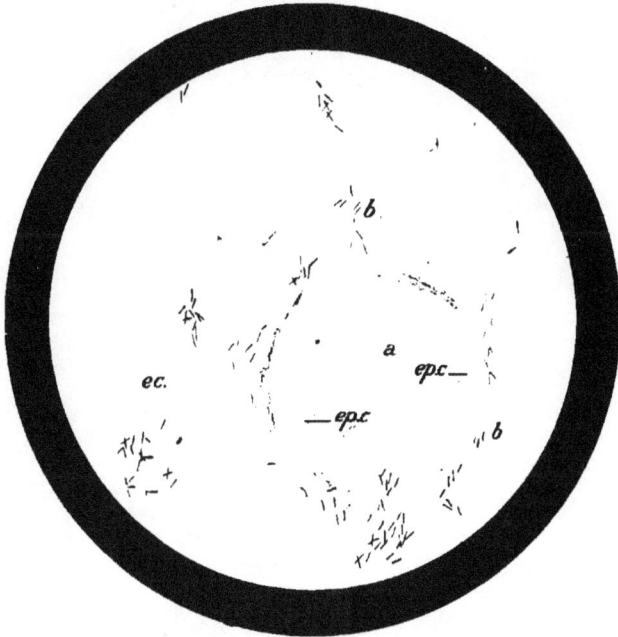

Tubercular Pulmonary Consumption. Microscopic Specimen from Sputa containing (b) Bacilli, (a) Alveolus, (e c) Endothelioid Cells, (ep. c) Epithelial Cells, Lung Tissue stained with Methylin Blue—Bacilli with Gentian Violet.

MOUNTED AND DRAWN BY H. MACDONALD M.D. N.Y.

right lung is probably the one most often primarily affected, the right bronchial tube, as so often stated, being larger than the left. But the reasons why the apices are affected primarily appear to be: (1) injury to the vessels here from imperfect circulation and stagnation of blood, exposure, and strain from coughing or lifting, giving rise to exudation; (2) limited respiratory motion with imperfect ventilation and non-absorption of exudations—conditions all of which favor the lodgment of bacilli.

1. CATARRHAL PHTHISIS.

Catarrhal phthisis (caseous phthisis, caseous infiltration, broncho-pneumonic phthisis, bronchitic phthisis, tuberculous lobular pneumonia) is one of the two varieties of chronic phthisis. It may assume various forms, but as commonly met with is usually divided into three stages:

1. A stage of incomplete consolidation of lung-tissue with more or less localized secondary bronchitis, usually of the smaller tubes, lobular pneumonia, or circumscribed pleurisy, or all three.

2. A stage of complete consolidation of lung-tissue, commencing softening, and one or all of the attending complications mentioned.

3. The breaking down of tissue and the formation of cavities. All the complications mentioned are more frequently present in this stage.

Inspection. First Stage.—This may give negative results. Usually, however, even in this stage, expansion of the chest over the affected portion may not be as well marked as on the healthy side, especially on deep inspiration. There may also be some slight flattening over the affected part, and when at the apex, the clavicle on the affected side may be more prominent than the other. The respiratory movements are some-

8

what increased in frequency, and the apex-beat of the heart will be more rapid than in health, but in the normal position, unless displaced by some complication. Though the appetite is usually poor, even in this stage, emaciation is not yet usually noticeable.

Palpation. First Stage.—The vocal fremitus is usually exaggerated (slightly increased) over the affected part, since the latter now conducts the voice sound better than in health. As solidification becomes complete so does the fremitus increase, owing to obliteration of air-cells, by which the affected part becomes a more homogeneous medium for the conduction of sound, instead of refracting it as in health. It should not be forgotten, however, that the vocal fremitus is normally exaggerated under the patient's right clavicle in health, on account of proximity to the large right primitive bronchial tube, by which the voice sound is conveyed with considerable force to that region. The skin is usually unnaturally warm and dry. The apex-beat of the heart is felt to be more rapid than in health, as also the pulse.

Percussion. First Stage.—There is the quality of slight dullness on percussion over the affected part, with consequent slight raise in the pitch, shortening of the duration, and diminution of the intensity (or amount) of the note in the sense of volume. But slight dullness is obtained in health in the right subclavicular region, on account of the right muscles being thicker than the left, and other reasons already stated (see Percussion of the Chest in Health. Chap. I.).

Myoidema (muscle swelling) is a knot that rises up from the surface of the chest, in some cases, after a sharp blow upon the latter immediately with the point of the finger. It may be observed now, but is more marked in the later stages, when emaciation is more noticeable. It was first observed by Stokes, of Dublin, about 1830, but was so named by Lawson Tait, of Birmingham (Dublin *Jour. Med. Science*, 1871, vol. 52, p. 316; It is obtained over the pectoral muscles chiefly, but sometimes also over the deltoid, and muscles of the back. It is due to muscular irritability, as may occur with emaciation from any cause, and was at first called by Schiff and others "idio-muscular contractility." It was thought to be pathognomonic of phthisis at one time, even in the absence of other signs, but at present it is regarded as of little value, and is observed to occur in other diseases, as typhoid fever, and diphtheria, and indeed I believe it may be obtained in some cases where there is perfect health.

Auscultation. First Stage.—The respiratory murmur over the affected part has now lost its purely vesicular quality, the vesicles or air-cells of the diseased part having been more or less obliterated by the presence of tubercles, causing pressure and local inflammation. The breathing is therefore a mixture of bronchial and vesicular or broncho-vesicular, as Flint termed it. As Da Costa says, however, it is better to call it vesiculo-bronchial, that being the order in which the two qualities occur. Inspiration is increased in intensity, in the sense of concentrated amount, being conducted with more force by the affected tissue than in health. It is less vesicular than in health, and slightly raised in pitch. It is barely finished, and hence is not quite continuous with expiration. The more complete the consolidation and the greater the area, the more un-finished and shorter will be inspiration. This, as well as bronchial breathing, therefore, is said to be divided. Expiration is as long as or longer than inspiration, it is as intense, the pitch is as high or higher, and the quality is more tubular. This vesiculo-bronchial murmur

FIG. 13.—Vesiculo-bronchial Breathing.

was formerly, and is now, by some termed rude, because it is more intense than normal. Others speak of it as harsh or rough breathing, all of which are unde-scriptive of the true characteristics. Either inspiration or expiration or both may be wavy, jerking, cog-wheeled, or interrupted, as one may choose to term it. This variation is thought by some to be due to the sudden passage of air through some point or points of stricture in the bronchial tubes, due to pressure from tubercles or obstruction from mucus. But it is also found in perfectly healthy lungs among some, for instance, who have palpitation of the heart, or are nervous from any cause. It is not at all uncommon among women, men addicted to the abuse of alcohol or tobacco, the choreic, and other nervous subjects. It is of great importance to remember that the normal vesiculo-bronchial breathing is heard in the right subclavicular region, owing to the proximity to the large right bronchial tube.

Pectorophony (vocal resonance), like the fremitus, is exaggerated (slightly increased) over the affected part.

Adventitious sounds are usually heard over the affected part in the first stage of phthisis, and it is by means of these, chiefly, that we are enabled to make an early diagnosis, especially if the seat of the disease be at the right apex. Along with tubercle there will be a localized capillary bronchitis, a spot of pneumonia, or perhaps pleuritis. For that reason we are almost sure to hear some subcrepitant, crepitant, or intra-pleural râles. Sometimes there may be a single click heard, during inspiration especially. It is called the mucous click, and is said to be due to the forcible separation of an agglutinated tube or overcoming mucous or other obstruction by air during inspiration. It is probably of intra-pleural origin sometimes.

Any of these localized adventitious sounds, in addition to the other physical signs mentioned, and the history of the case, would render it almost certain that the case was one of phthisis. The tubercle bacillus will rarely be found in the sputa at this early stage, even by an expert, because it will not usually have been liberated, or only bacillary germs may yet exist. But in case of doubt the sputa should be examined by an expert if possible.

Regarding the early diagnosis of the disease, it is seen from the physical signs presented, that it is more readily made when the disease affects the left apex, since all of the early physical signs, except the localized adventitious ones, already exist in the patient's right subclavicular region in health. *

Second Stage.—We have now to deal with complete solidification of lung tissue of variable extent, commencing softening, and local complications (p. 113).

---

* See also *Hæmoptysis*, p. 102. Unless caused by heart disease or injury, it is probably due to phthisis (Flint).

Inspection. Second Stage.—The expansion of the chest walls over the affected part is more markedly diminished on inspiration than in the first stage, and the difference between the respiratory movements of the unaffected and diseased sides is more noticeable. Flattening over the affected portion, due to retraction of the diseased lung tissue, is usually noticeable in this stage. If it occurs at the apex, the clavicle usually projects and is noticeably prominent. Respiratory movements are increased in frequency and rather shallow, and the apex-beat of the heart is usually quite visible, owing to emaciation, which now generally becomes noticeable. The apex-beat of the heart may be in the normal position, but is not infrequently displaced by traction due to pleuro-pericardial adhesion.

Palpation. Second Stage.—The vocal fremitus is usually noticeably increased on account of the better conducting power of the consolidated lung tissue. If, however, a large-sized bronchial tube leading to the affected part becomes obstructed, so that it ceases to convey the voice sound, the fremitus may be diminished, or even absent. Upon coughing and removing the plug, the fremitus at once returns. Thickened pleura or effusion intervening, would also cause the fremitus to be diminished or absent. The apex-beat of the heart may be felt to be abnormally frequent and feeble, as also the radial pulse. The surface of the body is usually felt to be dry and unnaturally warm.

Percussion. Second Stage.—Dullness on percussion is the rule, and this quality will be marked in proportion to the extent of consolidation. The more marked the dullness, the higher will be the pitch, the shorter the duration, and the less the intensity of the percussion note, in the sense of volume. It happens, however, once in a great while, that Skodaic resonance (tympanicity in connection with solidification or effusion) is obtained. This requires rather forcible percussion over solidification near a large bronchial tube, the trachea, an empty stomach or transverse colon, where the force of the percussion blow extends to them from the solidified tissue, which acts on the principle of a solid pleximeter, like the finger, or other solid percussion medium, and is consequently attended with a feeling of resistance. Even cracked-pot resonance, due to the concussion of air in such localities, is sometimes, though rarely, obtained. Over the unaffected parts the percussion resonance may be exaggerated, owing to the presence of a certain amount of vicarious emphysema, which, however, never becomes marked, as the volume of blood becomes much diminished in this wasting disease.

Myoidema is more noticeable in this stage than the first.

Auscultation.—The breathing over the consolidated lung tissue now becomes bronchial, all vesicular quality having disappeared with the air-cells. Both inspiration and expiration are tubular in quality and high pitched, expiration being the higher. Expiration is also as long as inspiration, or longer, and the two are separated, or divided, by inspiration being unfinished. This respiratory murmur as a whole is more intense than normal in the sense of concentrated amount, and expiration is more intense than inspiration. Both are somewhat short owing to the shallow breathing.

Pectorophony (vocal resonance) is usually markedly increased over the consolidated part, and becomes bronchophony (bronchial voice). Should the speech (articulate words) be heard also, it might be called bronchiloquy (instead of pectoriloquy, which also applies to cavities). The whispered voice would give whispering bronchophony or whispering bronchiloquy, as the case might be. Should a large tube leading to the part be obstructed, or there exist thickened pleura or effusion, the breath and voice sounds would be diminished or absent, like the fremitus under the same conditions. As some parts of the diseased lung will probably be in the first stage, or incomplete consolidation, the physical signs belonging to that stage may be looked for about the periphery of the seat of complete consolidation. Or both lungs may be affected with different stages. Localized adventitious sounds are also heard about the affected part in this stage. They are more marked usually than in the first stage, and consist of bronchial moist râles of various kinds, and sometimes intra-pleural râles. The crepitant râle, if present, would usually be obscured by other louder râles. Owing to commencing softening and breaking down of lung tissue, and perhaps pleuritic complications, certain crackling and crumpling râles, called indeterminate, may be heard sometimes.

Third Stage.—In this stage, cavities are formed. But in addition to cavities, there may be portions of lung tissue yet in conditions of complete and incomplete solidification, so that we may have all three stages present at once. Besides these, the usual secondary inflammations also exist, such as localized bronchitis

and pleuritis. We therefore usually have a complicated pathological condition to deal with, and hence,
although rules usually hold good, we are not to be
surprised at variations and exceptions.

Inspection. Third Stage.—The conditions already
observed in the second stage will be present, but usually to a more marked degree. There are emaciation,
rapid and shallow respiration, usually marked depression or flattening of the chest walls over the affected
part, diminution of respiratory movements, especially
on the affected side, and prominence of the superficial
veins. When the apex is the seat of the disease, the
supra- and infra-clavicular spaces are markedly depressed, as a rule, with corresponding prominence of
the clavicle. Why are the superficial veins usually so
prominent? Not so much on account of obstruction
in the pulmonary circulation, because, as already stated,
the volume of the blood in this wasting disease is much
diminished, nor is the obstruction to the pulmonary
circulation so great as it is in general hypertrophous
(large-lunged) emphysema. Yet in the latter disease
the superficial veins are not so prominent as in the third
or even second, stage of pulmonary consumption. The
true reason seems to be, that in consumption, as emaciation progresses, the vessels become more and more
prominent, owing to the absorption of the fat around
them. The impulse of the heart, which beats rapidly,
is usually visible by reason of emaciation of the chest
walls. The apex-beat may be in the normal position, but
it is not unfrequently displaced, sometimes markedly,
not usually from pressure, however, but from traction,
due to pleuro-pericardial adhesion and retraction of

the lung. This is especially the case when the left lung is affected. Walshe mentions a case where the apex-beat was, found under the left clavicle. It may be in either axillary line.

Occasionally inspection yields almost negative results, even in this stage, the disease having become temporarily arrested, and the patient improved in general health for the time being. (See *Hæmoptysis*, p. 102.)

Palpation. Third Stage.—The vocal fremitus is increased over the affected part, as a rule, owing to the consolidated lung tissue near the cavity. But in rare instances the increase may be due to re-enforcement of the voice by echo in the cavity, a phenomenon of the consonance of sounds. All the conditions necessary for this, however, are rarely present at the same time. Sometimes, instead of being increased, the fremitus may be diminished, or even absent, for reasons already enumerated, such as obstruction in a large bronchial tube leading to the affected part, thickened pleura, effusion, or intervening healthy lung tissue, in case of a very small, deep-seated cavity. Rhonchal fremitus due to gurgles, or other adventitious sounds, is sometimes felt, the thinness of the chest walls, due to emaciation, favoring its production. The apex-beat of the heart is usually felt, though feeble and rapid. The skin is usually hot and dry, and the radial pulse, like the heart, of course, frequent and feeble.

Percussion. Third Stage.—Dullness on percussion is the rule, especially gentle percussion, the consolidation around the cavity which contains more or less fluid generally, and the condition of the pleura being favorable for its production. There are, however, exceptions.

(1) Cracked-pot resonance is sometimes obtained. It is not known why it was ever termed cracked-pot, for it does not sound like percussing a cracked pot. The sound is in most cases due to the sudden forcing of air out of a cavity through a somewhat small opening communicating with a bronchial tube. The patient's mouth should be open, otherwise the air will not be driven out with sufficient force to produce the sound. It requires a cavity, superficially located, at the upper part of the thorax, with tense but yielding walls, and a sharp percussion blow. It is best obtained in emaciated women, as their chest walls are more yielding than those of men. The sound is exactly imitated, not by striking a cracked pot, but by clasping the two hands together, so that the palms form a cavity, and striking the back of one hand or the other on the knee with considerable force. Children do this to imitate the sound of money in their hands, and for this reason it is sometimes called the money chink resonance. But as it sounds very little more like the chink of money than it does like striking a cracked pot, and as the latter has the sanction of age as well as the name of Laennec associated with it, there appears to be no good reason as yet for changing it. Instead of being caused by the sudden escape of air from a cavity, it appears, in rare instances, to be due to concussion of air in some hollow viscus. Cracked-pot resonance is not constant, as the conditions for its production rapidly change, such as plugging of the opening with mucus, filling of the cavity with fluid, and so on. In the first case coughing may dislodge the plug, and the sound will reappear.

(2) Tympanitic resonance (drum sound) may be obtained by percussing rather forcibly over a large closed cavity filled with air, and having tense and rather unyielding walls. The pitch and other properties of the note will vary with the size of the cavity and the tension of its walls.

(3) Amphoric resonance (jug sound) is obtained by percussing, rather forcibly, over a cavity situated superficially at the upper part of the thorax, and having hard, smooth, and unyielding walls, empty or nearly so, and communicating freely with a bronchial tube—in other words a jug-like cavity (latin word *amphora*, signifying a jug, jar, or bottle) which acts as a resonant chamber. This amphoric percussion resonance may also be called metallic, the former having reference to the shape of the cavity, the latter to the metallic quality of the sound. In order to produce it the patient's mouth should be open during percussion, otherwise the sound would not be produced any more than by striking an empty jug with its mouth tightly corked. It is imitated exactly by percussing over the buccal cavity with the cheeks and lips tense, and the mouth partly open. The pitch varies in different cases, owing to different causes, as already explained, but the quality remains the same.

Gairdner and Finlayson found that over such cavities, and especially in pneumo-hydrothorax, a marked amphoric or metallic resonance is produced by means of percussing with one coin on another coin (fifty-cent piece for instance), as a pleximeter, anteriorly, while the auscultator listened posteriorly—the so-called bell-metal test. These cavities are usually situated high up toward the clavicle.

(4) Flatness on percussion may be obtained over a large superficial cavity filled with fluid.

(5) Normal resonance may be obtained on gentle percussion over a deep-seated small cavity, if healthy lung tissue intervene. If the examination be made posteriorly and laterally, however, as well as in front, the cavity will usually be found.

Myoidema, as already described, is usually noticeable in this stage, and especially on the front of the chest. For producing it, a sharp blow, delivered directly with the point of the finger, is required. It is of little value as a physical sign.

Auscultation. Third Stage.—As a rule, cavernous breathing is heard over the cavity. That is, the respi-

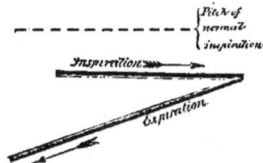

FIG. 14.—Diagram Showing Cavernous Breathing.

ratory murmur is blowing in quality both on inspiration and expiration, giving the idea, as Flint correctly states, of air passing in and out of a hollow space. The pitch is usually lower than normal, expiration being lower, and as long as, or longer, than inspiration, and the two continuous. But the pitch, of course, varies with the size of the cavity, the tension of its walls, and so on. A large cavity, other things equal, would give a lower-pitched cavernous respiratory murmur than a small one. Exceptions to this general rule are: (1) it

9

may be caverno-bronchial. Here we have a mixture, in various proportions, of cavernous and bronchial breathing, owing to the presence of a cavity, perhaps rather small, and a good deal of consolidated lung tissue at the same time. Flint was the first to observe this variety of abnormal respiratory murmur, and termed it broncho-cavernous. That it does exist in rare instances cannot be denied. But in the few cases where I have noticed it, the bronchial element came last, so that I have usually described it as caverno-bronchial breathing. (2) Amphoric respiration, or jug breathing, is the respiratory murmur heard over a jug-shaped cavity, and sounds like blowing across, or into, the mouth of an empty jug or bottle. The cavity must have hard, smooth walls, and a large mouth, and be empty or contain but little fluid. Amphoric breathing is heard over the same kind of a cavity that yields amphoric resonance on percussion, although the two may not always exist at the same time, in any given case.

The *Vocal Resonance* is best appreciated when the patient whispers. As a rule, cavernous whisper (whispering cavernophony) is heard over a cavity. That is, a whisper of blowing quality, and rather low pitch, differing in the latter respect from bronchial whisper, which must be high pitched, as we have seen. The pitch, usually low, varies, however, with the size of the cavity, the amount of solidified tissue, and so on.

Pectoriloquy.—In some cases, instead of cavernous whisper, the articulate words may be heard, and then it is usually called whispering pectoriloquy. Now pectoriloquy means literally chest speech, and was first applied by Laennec to the vocal resonance heard over cavities when the articulate words were heard. Not that one is able to carry on a conversation with a patient through a cavity, but when only a distinct syllable or two may be heard in some simple phrase. But pectoriloquy in the same sense is sometimes heard over consolidated lung tissue, or even in health, as well as over a cavity. Pectoriloquy, therefore, is a general term for expressing a class of vocal resonance, of which there are several varieties. Bronchiloquy may therefore be used, it seems to me, to express that variety of pectoriloquy sometimes heard over consolidated lung tissue, and caverniloquy when heard over a cavity. Indeed, Guttmann maintains that bronchophony and pectoriloquy are practically identical. So they are, if we mean that it is possible sometimes to obtain bronchial pectoriloquy. But bronchophony and bronchiloquy are both necessarily high pitched, as we have seen (Lobar Pneumonia, second stage), whereas Flint's whispering pectoriloquy (caverniloquy), as well as cavernous whisper (whispering cavernophony) and cavernous breathing, are usually low pitched and blowing in quality, instead of being tubular. Amphorophony and amphoriloquy (amphoric voice and speech), whispered or otherwise, are sometimes heard over a cavity, and sound like speaking out loud, or whispering, into the mouth of an empty jug or bottle. Amphoric echo is simply the reverberation of the voice, cough, or breath sound, in such a cavity. The same sort of cavity is required for amphorophony or amphoriloquy as in the case of amphoric resonance on percussion, amphoric respiration, and metallic or amphoric tinkle.

Ægophony (goat's voice) may also be heard in some

rare cases, but this will be fully considered when speaking of pleurisy. It is nothing more than distant, tremulous bronchophony; distant on account of some intervening medium, and tremulous from the presence of fluid thrown into vibration by the voice. Should the articulate words be heard, it would constitute what may be termed tremulous bronchiloquy.

Adventitous Sounds. Third Stage.—As a rule gurgles are heard. These are coarse, bubbling râles, made in the cavity, more particularly on inspiration, since the air enters the cavity with more force than it leaves it. For this reason the fluid in the cavity is stirred up more on inspiration than expiration. If the cavity be small, the gurgles may be fine. They therefore vary with the size of the cavity and the amount and consistency of the contained fluid. Their pitch also depends upon the amount of solidified tissue, which condition, as already stated, causes them to be raised in pitch. Whenever râles are mentioned as being gurgles, however, it is understood that a cavity is present. Metallic tinkle is sometimes heard. Laennec compared this sound to the dropping of water from a height into a metallic basin, and termed it metallic tinkle. It is also compared, by some, to the dropping of a pin into an empty bottle. But the idea of dropping anything is wrong, since metallic tinkle is probably never, in any case, produced by the dropping of anything. In the first place, the fluid in a cavity is generally slimy, and will slide down the walls of a cavity rather than drip off. In the second place, the cavity is too small to hear fluid drop from the vault into fluid underneath, even if it did drop instead

PLATE V.

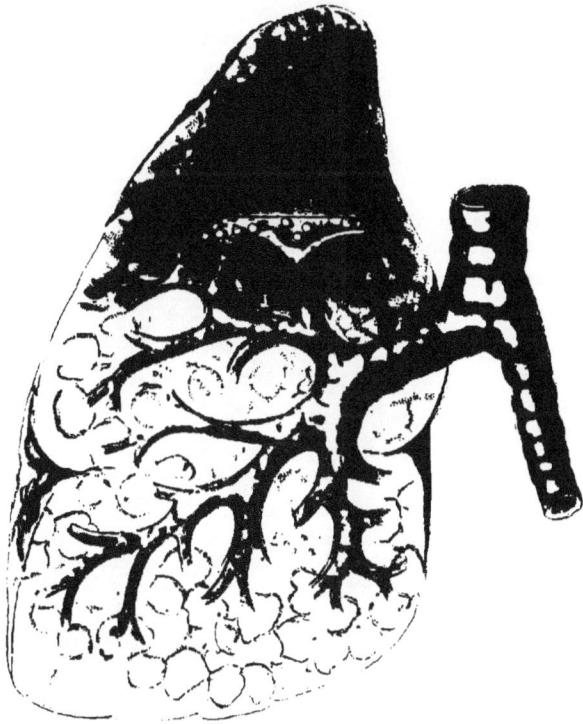

Phthisis Pulmonalis; Third Stage, showing Cavity with Gurgles, Upper Lobe of Right Lung.

BY H. MACDONALD M.D. N.Y.

of sliding. But this distance would further be diminished by the tenacity of the fluid, which would allow it to string down part of the way before it broke off. This dripping theory, therefore, it seems to me, must be abandoned, especially in so small a cavity as that usually found in the lungs. It is, in fact, produced by the bursting of bubbles of a small amount of viscid fluid in a cavity containing air, and which is a resonant chamber, the same in which we get amphoric breathing and amphoric voice. Metallic tinkle is usually heard most distinctly on inspiration. It may be heard while the patient holds the breath, if a bubble previously formed now bursts. Not all the bubbles will produce metallic tinkle, but only those that burst with a sound in consonance with the echo of the cavity. Metallic tinkle, in other words, is purely a phenomenon of the consonance of sounds. It is not constant, but appears and disappears in the same patient, according to the conditions necessary for its production. It is simply the occasional musical echo of some adventitious sound (a gurgle, usually) in a reverberating cavity possessing the qualities of amphoricity.

Bronchial râles of various kinds are usually present also, owing to the localized secondary bronchitis. They are usually of the moist class, and may be subcrepitant, submucous, or mucous. The subcrepitant and submucous are most common. Intra-pleural râles and friction sounds may also be present. In addition to these there may be, and often are, indeterminate râles; that is to say, certain crackling and crumpling adventitious sounds, partly dry and partly moist, and due to the breaking down of tissue. The crepitant râles, if pres-

9

ent, would probably be obscured by other and louder sounds.

Differential Diagnosis.—This rests chiefly upon the physical signs of more or less consolidated lung tissue and cavities, with secondary local inflammation, affecting the upper part of the lungs preferably. Bronchitis is a general, bilateral disease and gives no signs of solidification. Cancer of the lungs does this, but the dullness often extends across the median line, owing to co-incident enlargement of mediastinal tumors, and the sputa are currant jelly. There is also the want of hectic fever in cancer. Old pleurisy, with effusions, may be mistaken for phthisis, especially if the examination be confined to the apices of the lungs. But flatness on percussion, changing with position of the patient, due to effusion, will be found at the lower part of the thorax in pleurisy. Absence of respiratory murmur over the seat of disease, as well as other signs of effusion, will also be noticed (see Pleurisy). In cases of doubt the hypodermic syringe may be used. Only the first stage of phthisis offers room for doubt about the diagnosis as a rule. Then the signs of incomplete consolidation, so fully described, together with any of the localized adventitious sounds will usually lead to a correct conclusion. The discovery of the bacillus under the microscope would be conclusive, but the general practitioner need not expect to be able to do this, since he has generally neither the time, the proper instruments, nor the experience.

2. FIBROID PHTHISIS.

Inspection.—There is retraction over the affected part, which may be at the upper region of the thorax on the right side, for reasons already mentioned. The shoulder is usually lowered on the affected side, but not so much as in pleurisy with retraction, and the respiratory movements on that side are diminished or even absent, especially if pleuritic adhesions be present also, as is often the case. If the other lung be unaffected it becomes more or less enlarged, as also healthy portions of the affected lung, from vicarious emphysema. The apex beat of the heart is often displaced, sometimes greatly so. When the right lung is affected, the apex beat of the heart may sometimes be found in the right axilla. This is due to great shrinkage of the right lung and emphysematous enlargement of the left, and often pleuro-pericardial adhesions add to it by traction. Should the left lung be affected, the apex beat may be found displaced as far as the left axillary line. The shrinkage of the thorax is usually horizontal, with decrease in the antero-posterior as well as other horizontal diameters. But it also sometimes settles down, especially the upper part, with decrease of vertical diameter. In these cases the apex of the heart, if not otherwise displaced, will be found between the fourth and fifth ribs instead of the fifth and sixth. The intercostal spaces are sometimes much narrowed.

Palpation.—The vocal fremitus is increased over the indurated tissue, unless there be obstruction in the

tubes from stricture or other cause, or interruption due
to thickened pleura, when it may be diminished or
even absent. The fremitus, therefore, though increased
usually in proportion to the consolidation, may vary
for different cases. The apex beat may be felt out of
its normal position or not, according as the heart is
displaced or not.

Percussion.—There is usually marked dulness on
percussion over the indurated lung and exaggerated
resonance over the emphysematous portions. Owing
to induration of lung-tissue, narrowing of intercostal
spaces, and hardening of the ribs, the dulness is sometimes termed wooden, and there is great feeling of resistance to the finger on palpatory percussion, as Piorry
first noticed. Hence, as often before remarked, one of
the advantages of palpatory percussion over all other
methods. Sometimes, if there be a large bronchiectatic
cavity, tympanitic, cracked-pot, or other resonance indicative of a cavity may be obtained (pp. 123–125). The
line of hepatic dullness will be unusually high up if
the vertical diameter of the thorax be much shortened
(pp. 136, 148).

Auscultation.—Bronchial breathing and bronchophony will be heard over the contracted lung unless
convection of the sounds of the respiratory murmur
and voice by the bronchial tubes into the part be obstructed by stricture of the bronchi, or plugging of
them with viscid mucus, or other cause. Or else there
may be interruption of these sounds from thickened
pleura. In either case the respiratory murmur and whispered voice will be weak or suppressed over the corresponding area, but the voice, if uttered loud enough, will

surely be heard, though the resulting bronchophony
may be distant or weak. Should bronchiectatic cavity
be present, as generally occurs in the course of the disease, the signs would indicate it as already described
(pp. 125–129). ·No cardiac murmur, unless it already
exists, necessarily accompanies displacement of the
heart. Owing to obstruction to the pulmonary circulation the right ventricle becomes somewhat enlarged
with accentuation of the second sound over the pulmonary (second left) interspace (pp. 188, 189). Owing
to diminution of the volume of blood in this wasting
disease, however, these signs will not be so well marked
as in general large-lunged emphysema, where there
is not only marked obstruction to the pulmonary circulation, but also a full volume of blood not infrequently, to be driven through the lungs.

Differential Diagnosis.—Unless this disease affects
both lungs it may be impossible to distinguish it from
chronic (interlobular, interstitial) pneumonia, which is
nearly always unilateral. In the pneumonia, patients
do not have so much fever, retain more strength, and
do not become so rapidly and markedly emaciated.
Moreover, they do not suffer from the hoarseness and
diarrhœa due to tubercular disease of the larynx and
intestines that frequently occur in fibroid phthisis.
The tubercle bacillus is not so abundant in fibroid
phthisis as in the catarrhal form, and its discovery,
though decisive, is often difficult and even impossible. Fibroid phthisis is said to affect the upper
lobes by preference, while the pneumonia is usually
basic.

Pleurisy with retraction may be difficult to differentiate. Indeed, both are not infrequently present, so that by shrinkage of the pulmonary interstitial connective tissue and compression by the contracting and enormously-thickened pleura, the lung may become as small as a man's fist. But in pleurisy there are no signs of cavities as in fibroid phthisis, and in the former there are diminution or absence of the respiratory murmur and fremitus, the very opposite being the case in consolidation of lung unless there be obstruction due to bronchial stricture and the like. The heart is not so much displaced in pleurisy, and in the latter disease the patient is also usually in better general condition and with less cough and expectoration.

In cancer of the lung, in addition to the cancerous cachexia, dullness over the affected part may extend across the median line, with inward pressure signs on the trachea and œsophagus due to cancerous enlargement of mediastinal lymphatic glands. The percussion dullness from solidified lung-tissue, on the contrary, never extends across the median line. The sputa in the two diseases would also differ.

Compression atelectasis from pleuritic effusion, aneurismal or other tumors, deformities, and the like (p. 60), may be readily differentiated by careful examination and ascertaining the cause of the atelectasis.

During the recent Berlin Medical Congress, Professor Ewald, of Berlin, expressed the opinion that chronic pneumonia is phthisis, the reason why it is chronic being due to the presence of the tubercle bacillus. According to Williams, the tubercle bacillus is nearly always found even in the so-called potter's and knife-grinder's lung and similar diseases.

CHAPTER IV.

Diseases in which the breath and voice sounds are interrupted in
their transmission to the chest walls, with consequent diminution, or absence, of vocal fremitus, respiratory murmur, and
pectorophony.—Diseases of the pleuræ.—Thickened pleuræ.—
Fluid or air, or both, in the pleural cavities.

PLEURISY.

PLEURISY, or pleuritis, is inflammation of one or both
pleuræ. These, consisting of two layers, visceral and
costal, form a closed sac on each side. These sacs do
not communicate, but approach each other closely behind the median line of the sternum, from a point corresponding with the upper borders of the second costal
cartilages, to the upper borders of the fourth, where
they diverge. The left pleura turns obliquely downward and outward, but within the left nipple to the
fifth cartilage, leaving a portion of the heart uncovered
—the superficial area of cardiac dullness; thence inward to the upper border of the sixth, where it turns
outward again to the lower border of the sixth costal
cartilage on the left mamillary line; thence to the lower
border of the eighth, on the left axillary line; the
ninth rib on the scapular line (or line let fall perpendicularly from the inferior angle of the scapula), and
the tenth rib on the vertebral line (close to the spinal
column).

The right pleura continues down behind the median
line of the sternum to the upper border of the sixth

costal cartilage where it turns off nearly at right angles along the upper border of the liver. On the right axillary line the right pleura reaches down to the upper border of the eighth rib, on the scapular line the ninth rib, and on the vertebral line the tenth rib.

The following is a summary of the lower limits of the pleuræ:

Left Pleura.—On sternal line, upper border of fourth costal cartilage. On mamillary line, lower border of sixth. On axillary line, lower border of eighth rib. On scapular line, ninth rib. On vertebral line, tenth rib.

Right Pleura.—On sternal line, upper border of sixth costal cartilage. On mamillary line, the same. On axillary line, upper border of eighth rib. On scapular line, the ninth rib. On vertebral line, tenth rib.

These limits vary somewhat with deep inspiration, also in emphysema, effusion of fluids into the pleural sacs, pneumothorax, and the like.

The chief function of the pleuræ is to furnish a small amount of lubricating material within the sacs, so that the layers can easily, and without noise, glide on each other, and thus facilitate respiratory movements. In this respect they resemble the synovial membrane of a large joint. And as anchylosis affects a joint, so are pleuritic adhesions injurious in proportion as they are extensive, and restrict the movements of respiration.

Pleurisy may be divided into two classes: (1) dry pleurisy, and (2) pleurisy with effusion. Of each of these classes there are three varieties, (1) acute, (2) subacute, and (3) chronic.

Dry pleurisy, circumscribed pleurisy, pleurisy with

scant fibrinous exudation, or pleurisy without effusion, is a very common affection and often escapes
notice. It is usually this class of pleurisy that is found
in connection with pulmonary consumption, where the
inflammation crops out here and there in spots on the
periphery of the lung. Or it may be secondary to extension of inflammation from cancer or other neoplastic growths. It may be caused by fracture of a rib or
other surgical injury, or it may be due to exposure
to a draught of cold, and, finally, it may come on without any known cause. In idiopathic dry pleurisy the
physical signs are few. Usually there is jerking respiration or a catch in the breath, as it is called, owing to
the sharp pain in the side. This pain is usually near
one nipple or the other. On palpation the fremitus is
unchanged, owing to the small extent of the affection,
and the slight pathological change produced. This consists in a small spot of thickening or roughening of the
pleura. Sometimes if there be scant fibrous exudation,
a point of adhesion may result. The pulse is not ac-
celerated and there is no fever. The percussion resonance is normal. On auscultation a slight pleuritic
friction or crepitation may be heard, especially on inspiration. In a few days, however, this may have disappeared. When no adventitious sound is present it
might be a question as to whether the affection was dry
pleurisy, pleurodynia or intercostal neuralgia. But dry
pleurisy invariably has only one point of pain, which
the patient can locate with the point of the finger. In
intercostal neuralgia there are usually three points of
tenderness; near the spinal column, at the anterior extremity of the nerve, and about its middle. Besides

this the patient will probably give the history of previous attacks of neuralgia in other localities as well as intercostal. In pleurodynia, or myalgia, there is extreme tenderness in the muscles of the side. The patient shrinks from touch, and there is pain on motion. Lumbago, or other muscular rheumatism, is often associated.

Pleurisy with effusion is inflammation of the pleura, attended with sero-fibrinous effusion into the sac as the result of interstitial exudation. It may be primary or secondary, and either of these may be acute, subacute or chronic. The latter disease is sometimes described as empyema (pyothorax), but this will be described separately, and by chronic pleurisy with effusion will be meant old cases of pleurisy with sero-fibrinous, or serous, effusion.

Acute pleurisy with effusion, lasts about two weeks, usually, and. like acute lobar pneumonia, has three stages. Undoubtedly, in many cases, the exciting cause is exposure to cold and wet, when it is said to be primary, although some predisposing cause, notably lowered vitality, must also exist. According to Landouzy, of Paris, many of these cases of pleurisy *à frigore*, so-called, are merely symptomatic of latent tuberculosis and are not primary. This is probably true for Paris, but not all climates, as I have learned from personal experience, as well as by inquiry at the Brompton Hospital in London, and other hospitals.

Acute pleurisy with effusion may occur secondarily in the course of articular rheumatism, Bright's disease of the kidneys, scarlet fever, measles and pyæmic conditions.

The first stage of acute pleurisy with effusion is the dry stage of congestion, and lasts from a few hours to twenty-four hours, or, in some cases, even longer.

The second stage, that of effusion, lasts about five days on the average. The amount of liquid effused is not usually so great as in the subacute or chronic form, and is more fibrinous.

The third stage, that of absorption, usually begins by the eighth day, and in about two weeks, in an ordinary case, recovery may be said to be complete. The physical signs differ according to the stage of the disease.

First Stage.—The physical signs are similar to those of dry pleurisy already described, only more severe, as the disease under consideration is more extensive.

Inspection.—The patient lies on the affected side, since by this means the respiratory movements of that side would be restricted, while those of the unaffected side would be free. But with all precautions the respirations will be jerking, due to pain in the side. The respiratory movements are restricted on the affected side, somewhat exaggerated on the opposite side.

Palpation.—The vocal fremitus will be unchanged or very slightly diminished, owing to the slight thickening of the pleura from congestion. But in persons with thin chest walls a friction fremitus over the diseased pleura may sometimes be felt.

Percussion would yield normal resonance or slight dullness if the pleura be slightly thickened by the congestion.

Auscultation.—The respiratory murmur would be disturbed in its rhythm by the jerking due to pain, but its quality would be normal; so would the pectorophony (vocal resonance), but both would be somewhat diminished in intensity, due to slight thickening of the pleura from congestion.

A pleuritic. friction sound is usually heard on inspiration, expiration, or both. Honoré, of Paris, in 1819, was the first to call attention to this adventitious sound and to point out its true pathological significance.

All intra-pleural adventitious sounds are heard usually more distinctly on inspiration than on expiration, since the two layers of pleurae are brought closer and more forcibly together in the former act than in the latter. Those due to rubbing together of opposing roughened surfaces, and stretching of adhesions, are often heard only on inspiration, while intra-pleural râles, due to viscid exudative material, or perverted secretion, are often heard both on inspiration and expiration, and simulate very closely bronchial and other moist râles. Pleuritic friction may sound like the creaking or rubbing together of leather, or treading in deep, crusty, snow. Or it may be merely grazing or crepitating. Distinct pleuritic friction sounds are not infrequently interrupted, that is to say, there may be two or three consecutive friction sounds during inspiration, expiration, or both. If the chest walls are thin and the friction well marked, a localized friction fremitus may be felt on palpation as already stated.

Second Stage.—After effusion has taken place, we obtain the physical signs to be described in connection with chronic (subacute) pleurisy, with effusion (p. 142). They are chiefly those of interrupted transmission of sound, due to the presence of fluid in the pleural cavity acting as a partition or diaphragm, as well as flatness or marked dullness on percussion over the site of the effusion.

Third Stage.—After absorption has sufficiently progressed, the friction sound, which had disappeared during effusion, reappears—*frictio redux.* This also finally disappears in favorable cases, as the conditions producing it clear up, and the physical signs again become those of health.

An all-important question arises in connection with this disease. When must the chest be tapped? If life be threatened, owing to great effusion, tap at once. Do not draw off all the fluid, however, but only enough to allow the patient to breathe easily. But unless symptoms of suffocation are urgent, and they very rarely are, do not draw any of the fluid off before the end of the third week. Why? Because if the fluid be drawn off too early, and before active inflammation has subsided, adhesion of the two layers of the pleura will result, which would be one of the most unfortunate terminations that the disease could possibly have. The fluid in the sac keeps the two layers apart. To draw it off too early would not only thwart nature, but would be malpractice. According to Potain, of Paris, and others, only in rare and urgent cases should any fluid be withdrawn before the twenty-first day of the disease, and even then, only a part of it.

Subacute pleurisy with effusion usually results from a badly managed case of the acute variety, or in cases where the patient is in bad condition when attacked, or perhaps is tuberculous.

Chronic pluerisy with great effusion is described by many authors as subacute. It rarely follows the

acute form, but is usually chronic from the first, and
is not infrequently of tuberculous origin. The patient
may not complain of pain in the side at all. Usually
there is a hacking reflex cough, without much expectoration. The appetite is poor and emaciation
follows, attended sometimes by hectic fever. The patient applies to a physician, who, being satisfied to examine the apices of the lungs only, finds signs of consolidation from compression at one apex, and hastily
makes a diagnosis of phthisis simply. Had he taken
the trouble to examine the patient carefully down to
the waist, he would have found pleuritic effusion with
the following physical signs:

Inspection.—Usually there are bulging of the intercostal spaces over the seat of effusion, which is always
at the lower part of the thorax, displacement of the
apex of the heart laterally, in a direction opposite to
the seat of pressure from the effusion, and diminution
of respiratory movements on the affected side in proportion to the amount of effusion, with exaggerated
respiratory movements on the unaffected side. Collapse not infrequently follows removal and absorption
of the fluid and the other side now appears abnormally
large, partly from comparison with the collapsed side,
but partly also because the lung has become actually
larger in volume from vicarious emphysema.

Where the case is not of long standing, however, and the patient has chest walls covered with fat and muscle, the bulging
over the affected part may not be observable. Such a case applied to me in May, 1885. He was a fire-stoker, aged thirty-two.
The chest was perfect in outline. Upon my advice he applied to
the late Prof. E. D. Hudson, whom I then assisted at the New
York Polyclinic. Aspiration proved that the right pleural cavity

contained an abundance of sero-fibrinous fluid. In this case the effusion was due to tuberculosis, as *post-mortem* examination proved, nearly two years afterward.

Palpation.—The vocal fremitus is diminished or absent over the effusion, according to its amount. Above the level of the effusion, where the lung is in a condition of compressed atelectasis, the fremitus is increased. These signs may evidently change with position of the patient, unless the lung is fixed in its position by adhesions. The apex-beat of the heart may be felt out of its normal position. It sometimes happens that a spot of fremitus is felt somewhere over the seat of effusion. This is due to the telephoning of the voice to the spot along a string of adhesion, or else it may be extended by a rib, as in case of a heart murmur. Intercostal fluctuation may sometimes be felt.

Percussion.—Over the fluid there is flatness on percussion, unless the chest walls are thick, in which case marked dullness will be elicited, as the walls themselves will give out some resonance, like the thigh, or the deltoid muscle. But where the chest walls are thin and the fluid abundant, flatness, or absence of resonance, except what is obtained out of the finger as a pleximeter and the chest walls, is the rule. Besides, there is also a feeling of resistance to the finger on palpatory percussion in these cases. The upper line of the fluid, instead of being perfectly level, is, according to some, curved somewhat like the letter S, forming what is termed the curved line of Ellis. It is lowest in front, highest at the side, and averaged between the two posteriorly. This line changes with position of the patient unless prevented by adhesions. Over the compressed lung there is dullness on percussion, but not infrequently under the clavicle on the side of effusion we obtain tympanicity (Skodaic resonance). This occurs more frequently in front than behind. The true cause of this tympanicity was first explained by Skoda. The pulmonary vesicles being obliterated by compression, as well as the ultimate bronchial tubes, the compressed pulmonary periphery is gathered around a bundle of dilated bronchial tubes that form an irregular cavity, which yields tympanicity on percussion. There is exaggerated percussion resonance over the other lung, owing to its being in a state of vicarious emphysema. It sometimes happens that tympanicity, or even cracked-pot resonance, may be obtained over the seat of effusion. This is especially true in case of children. This is owing to the fact that the force of the percussion blow extends to the stomach, transverse colon or some neighboring viscus distended by air. Very gentle percussion, therefore, in children, is necessary to avoid tympanicity. The cracked-pot sound, if obtained, would be due to concussion of air in some hollow viscus, and not to its forcible escape.

Auscultation.—In listening over the seat of the effusion the respiratory murmur is diminished or, generally, absent. This is due to the fact that the lung is pressed away from the site of effusion, which not only does not conduct the breath and voice sounds but interrupts them. The pleural cavity, containing fluid, acts like a partition. The thicker the partition the less distinctly does sound pass through it. Sometimes, however, owing to a string of adhesion, bronchial breathing may be telephoned to a spot somewhere over the site of effusion,

or else it may be transmitted there along a rib. Above
the seat of effusion and over the compressed lung we
hear bronchial breathing. But occasionally cavernous
breathing is heard over the seat of percussion tympanicity. Over the other lung the breathing becomes exaggerated, owing to the extra work that lung has to do.

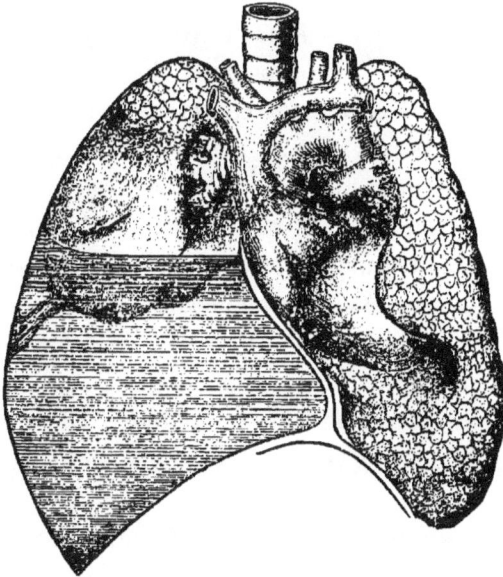

FIG. 15.—Diagram showing Pleurisy with Effusion, Compressed Lung, a String
of Adhesion from the Compressed Lung to Thoracic Wall, Dilated Bronchi and Displacement of Organs to the opposite side.

Pectorophony (vocal resonance) is very much diminished over the effusion, but is not entirely absent if the
voice of the patient is loud enough. The whispered
voice, like the respiratory murmur, is usually absent.
But in speaking out loud, distant or weak bronchophony is heard. If this distant bronchophony is tremulous, due to vibrating fluid, it is called ægophony
(goat's voice). Ægophony is usually obtained where

the effusion is not great, otherwise the bulk of fluid cannot be thrown into vibration. It is heard most commonly about the lower angle of the scapula, along the upper border of the fluid, when it is at that point. A thin stratum of encapsulated fluid may also be thrown into tremulous vibrations by the voice, thus causing ægophony.

Thickened Pleura.—It not infrequently happens that after effusion has disappeared the pleura remains thickened. Indeed it is of common occurrence to examine patients with only thickened pleura without effusion. In these cases the thickened pleura also acts as an interrupter to the transmission of sound. The respiratory murmur, pectorophony (vocal resonance), and fremitus are also diminished, and percussion dullness is noticeable, in proportion to the thickening. A string of pleuritic adhesion, however, may transmit sound on the principle of a telephone. But extensive pleuritic thickening is a different condition, which acts rather as a diaphragm, or partition, than a telephone.

Differential Diagnosis.—The diagnosis of idiopathic dry pleurisy, from pleurodynia and intercostal neuralgia, has already been considered. The diagnosis between acute pleurisy with effusion and acute lobar pneumonia is readily made, generally. In pleurisy with effusion there is interruption to the transmission of sound and the respiratory murmur, and the vocal resonance and fremitus over the affected part are diminished or absent, but markedly increased in pneumonia. The line of flatness or dullness in pleurisy often changes with position of the patient, but never in pneumonia. In cancer of the lung there are signs of consolidation, usually at the upper and front part of the lung with increased conducting power, the dullness often extending across the median line, owing to cancerous enlargement of mediastinal glands. In pleurisy with effusion, the affection is at the lower part of the thorax posteriorly, with interruption to transmission of sound and consequent diminution or absence of respiratory murmur and fremitus. The line of dullness in cancer does not change with position of the patient. In phthisis we also have signs of consolidation and increased conduction power of lung tissue, instead of the interruption of transmission of sound, as in pleurisy. In old cases of pleuritic effusion and compressed lung it is only carelessness should the physician confine his examination to the top of the lung and omit the lower part of the thorax. Thoracic aneurism has been mistaken for pleurisy with effusion, but such cases are very rare (see Aneurism).

Hydatid disease of the lungs may be mistaken for pleurisy with effusion. Such a case has aready been alluded to (see Hydatid Disease of the Lungs, p. 109). There were signs of effusion in the upper part of the right pleural cavity. The location excited suspicion, as pleurisy usually affects the lower part of the thorax. Under the microscope some of the aspirated fluid showed the presence of hooklets and scolices. In any case of doubt the hypodermic syringe may be used to test the presence or absence of fluid.

The differentiation of pleurisy with effusion from enlargement of the liver and spleen is sometimes difficult.

The upper limits of the liver have been already given.

being the same as the lower limits of the right lung
and pleura.   The lower limits of the liver correspond
to the free margin of the ribs in the right hypochondriac region, the tenth interspace on the right axillary line, and the eleventh rib on the scapular line, below which it is lost in the dorsal muscles.

The liver usually enlarges downward, but sometimes
upward, especially, according to Niemeyer, in case of
abscess or hydatid disease.   If the case be one of simple
enlargement of the liver upward, the line of percussion
dullness will extend up higher in front than behind,
the organ ascends and descends during respiration, the
respiratory sounds posteriorly though feeble, are not,
according to Da Costa, entirely absent, and the heart,
if displaced at all, is pushed upward.   The signs do
not change with position of the patient, as they often
(not always) do in pleurisy with effusion.   In the latter
disease, occurring on the right side, the liver may be
pushed down, giving dullness on percussion below the
ribs.   But, with the patient in the sitting or standing
position, the line of dullness will extend up as high
posteriorly as in front, and the liver does not move
up and down during respiration, but it is permanently
depressed.   Moreover, a small yielding interval may
be usually felt between the lower border of the ribs and
a normal liver that has been pushed down by fluid,
whereas no such space exists in case of enlargement of
the liver downward.

According to Niemeyer, the normal dullness of the
spleen is from the upper margin of the eleventh rib,
along the axillary line, to the ninth rib; anteriorly to
a line drawn from the anterior end of the eleventh rib

to the nipple; but posteriorly it cannot be defined from
the dullness of the left kidney, but its greatest thickness from before back, is about two inches. The spleen
usually enlarges downward and forward, and then
may be felt below the ribs in the abdominal cavity.
But according to Niemeyer and others it may extend
upward, rarely higher than the fifth rib, without going
below the margin of the ribs at all, even in cases of de-
cided enlargement. Intestines distended with gas, and
the like, push it upward and backward as in typhus,
but if it enlarges upward, as in intermittent fever, it
extends toward the axilla. The heart is pushed upward, and not sideways, the latter usually being the
case in large pleuritic effusion. The enlarged spleen
moves up and down perceptibly during respiration,
the level of dullness changing an inch sometimes
on respiratory percussion, accordingly as percussion
occurs at the end of a full inspiration or expiration. The area of dullness is less when the patient
lies on the right side in enlarged spleen, but it does
not otherwise change with position of the patient
as is often (not always) the case in pleuritic effusion. In enlarged spleen, the respiratory sounds are
feeble over the part, but, according to Da Costa, the
vocal vibrations are mostly unimpaired, whereas they
are usually absent in effusion. The rational signs of
the case and previous history, with reference to pleurisy, intermittent fever, and the like, will aid in arriving
at a correct diagnosis. Lastly the hypodermic syringe
may be used to test the presence or absence of fluid in
the pleural cavity. Extensive plastic (adhesive) pleurisy
without effusion would, among other points of differ-

ence, cause no fullness of the chest-walls, like enlarged
liver or spleen, and might even produce contraction.

In case of effusion into the pleural cavity it may become necessary to perform the operation of paracentesis thoracis (thoracentesis, tapping, or aspirating the
chest). This may be done by means of Potain's aspirator, or better still, perhaps, by Dieulafoy's instrument. I have had a guard, regulated by a screw, fitted

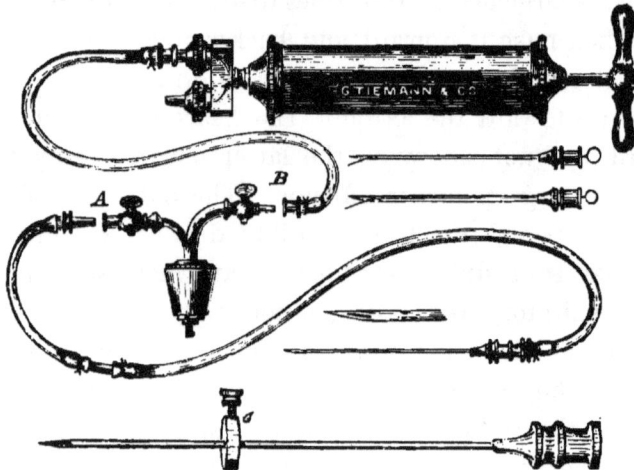

Fig. 16.— Potain's Aspirator, and Needle with Movable Guard G.

to the needles used by myself. By this means the
needle cannot accidentally penetrate any deeper than
is necessary to draw off the fluid.

Where shall we puncture? According to Bowditch, find the inferior limit of the sound lung behind and tap two inches higher than this on the
pleuritic side, at a point in a line let fall perpendicularly from the (inferior) angle of the scapula.
Push in the intercostal space here with the point of
the finger, and plunge in the needle quickly and firmly

at the depressed part, in order to get through false membranes. If no fluid be obtained here, puncture a little higher up and further toward the axillary line. Bowditch's point of puncture has the advantage over all others of allowing the pleural sac to be entered at the most dependent point, with less risk of perforating the diaphragm and causing peritonitis, though it possesses the disadvantage of obliging the patient to be in a sitting posture, if that be any, instead of the recumbent, during the operation.

Fraentzel, of Berlin, prefers a point half way between the mammillary and axillary lines in the fifth interspace on the left side, and in the fourth on the right, the patient being in the recumbent position. Aufrecht, of Magdeburg, prefers a point on the axillary line in the fourth interspace on both sides, the patient being recumbent. Puncturing the diaphragm with resulting fatal peritonitis has occurred in this operation by penetrating the chest-walls too low down. In general, the puncture should be made on the axillary line, between the fifth and sixth ribs on the left side, and between the fourth and fifth ribs on the right side to avoid the liver.

In what cases, speaking of adults, shall the operation be performed, and how much fluid be withdrawn? According to Anstie (Reynold's System of Medicine) it should be done (1) in all cases where the fluid fills one pleura and begins to compress the other lung; (2) in all double pleurisies where the total fluid would about fill one pleural cavity; (3) in all cases of large amount of effusion where there have been one or more fits of orthopnœa; (4) in all cases where the fluid is purulent, and (5) in all cases where fluid, occupying at least half

of a pleural cavity, has existed for a month and shows
no signs of being absorbed.

Neither Bowditch, Murchison, nor Anstie recommend
to withdraw all the fluid, but only so much as will
substantially relieve the mechanical distress caused by
pressure. In most cases that is all that is necessary to
excite the natural process of absorption. They all stop
the withdrawal of fluid the moment the patient begins
to complain of constricting pain in the chest or epigastrium.

Fraentzel, of Berlin, operates usually when the fluid
reaches up to the third rib, and œdema of the other lung
is threatened. He takes out about fifty ounces of fluid
and repeats the operation in three to five days when
necessary.

Aufrecht recommends (1) that a trial puncture by
Pravaz's syringe be made before operating. The ordinary hypodermic syringe with a perfectly fitting piston
and carefully adjusted long needle (disinfected) is
sufficient. (2) If there is reason to believe that less than
forty-five ounces of fluid will be discharged, the operation had better be usually left undone. (3) Not more
than eighty ounces of fluid should be evacuated. (4)
The thoracentesis should not be repeated unless a vital
indication demand it. (5) The operation of drawing
off part of the fluid should be performed in all cases
of large effusions as soon as the patient is seen, irre-
spective of the degree of temperature. Other authors
draw off the fluid at an earlier date than any of those
mentioned, and do not hesitate to withdraw it all.

In case of empyema (pyothorax) a permanent drainage tube, securely fastened to prevent its dropping into

the pleural cavity, promises better results than repeated aspiration. The tube may be introduced through a free opening as low down as the seventh intercostal space on the axillary line, whereas the trocar (needle) should not be pushed into the pleural cavity lower down than already mentioned.

HYDROTHORAX.

Hydrothorax is a dropsical and non-inflammatory affection, in which there is fluid in both pleural cavities. It is the result of a serous transudation and not of an inflammatory exudation, and is usually associated with general dropsy from some cause. Hydroperitoneum and hydro-pericardium may exist at the same time. The physical signs are similar to those of pleurisy with effusion, the later being generally unilateral, however, and hydrothorax bilateral. For this reason the heart is not noticeably displaced in hydrothorax, unless it be pushed down. Hydrothorax is, of course, unattended by friction sounds.

EMPYEMA.

Empyema, pyothorax, or suppurative pleurisy, is a disease characterized by pus in the pleural cavity. It may be due to traumatism, or an abscess opening into the pleural cavity from the liver, abdomen, the chest walls, or the lung. When it occurs without any of these causes it is probably due to some constitutional vice or to exhaustion of vitality. But why pleuritis should sometimes result in sero-fibrinous, and at other times in purulent effusion, is not exactly known.

The physical signs are almost identical with those of

pleurisy with non-purulent effusion. There is usually more emaciation in empyema, and the signs generally are more grave. But the only means of making a positive diagnosis is by withdrawing some of the fluid with the hypodermic syringe, or other aspirating instrument.

HÆMOTHORAX.

Hæmothorax, or blood in the pleural cavity, may be due to traumatic causes, or it may result from cancer of the pleura or rupture of aneurism into the pleural sac. Very rarely it is caused by the withdrawal of fluid from the pleural cavity, causing rupture of vessels by the sudden removal of the pressure to which they had become accustomed. Wounding the intercostal artery by the aspirating needle may sometimes produce it. The physical signs of blood in the pleural cavity are similar to those of pleurisy with effusion. Neither in hydrothorax nor hæmothorax are there friction sounds. The former is usually bilateral, hæmothorax unilateral. In hæmothorax, also, the symptoms are sudden and urgent, whereas hydrothorax is always insidious.

PNEUMOTHORAX.

Pneumothorax is a disease in which there is air in the pleural cavity. Generally it is unilateral. It may be due to traumatic causes, such as penetrating wounds of the thorax, injury to a lung from the end of a fractured rib, and the like. Or it may be due to openings into the pleural cavity from rupture or ulceration of the stomach or œsophagus, and from the lungs in the course of empyema, abscess, or hydatid disease. According to Walshe, however, ninety per cent. of all cases of pneumothorax are caused by the escape of air from the lungs into the pleural cavity, due to breaking down of tubercle. It is very doubtful if gas ever originates spontaneously in a closed pleural cavity. The physical signs are as follows:

Inspection.— Dyspnœa and anxious countenance are usually noticeable. Sometimes there is more or less cyanosis. Bulging and want of respiratory movement on the affected side with displacement of the heart in the opposite direction, are marked in proportion to the amount of air in the pleural cavity. Respiratory movements on the unaffected side are exaggerated. If the opening be valvular, so that air enters the pleural sac without escaping, the dyspnœa becomes extreme, and all the signs are marked.

Palpation.—The vocal fremitus is diminished or absent, according to the amount of air in the pleural cavity. The heart may be felt displaced and beating rapidly.

Percussion.—Over the affected side there is tympanitic resonance on percussion. The pitch will be high or low according to the volume of air in the pleural cavity and the tension of the chest walls. Over the other lung, exaggerated resonance is due to the extra work it is doing, and the state of vicarious emphysema.

Auscultation.—The respiratory murmur over the affected side is usually diminished or absent, unless there are string-like adhesions.

The vocal resonance is also diminished, and sometimes, according to Walshe, has a metallic (amphoric) quality.

The tympanicity on percussion at once distinguishes it from fluid in the pleural cavity from any cause. The diagnosis of pneumothorax from emphysema, has already been considered. (See Emphysema.)

PNEUMO-HYDROTHORAX.

Pneumo-hydrothorax, or hydro-pneumothorax, as the name indicates, signifies air and fluid both in the pleural cavity. According to some authors, the fluid is always purulent, and they describe it as pneumo-pyothorax, or pyo-pneumothorax.

As pneumothorax is always followed by inflammation and effusion into the pleural cavity within a few hours, or a day or two at most, the etiology of one disease applies also to the other.

Inspection.—Dyspnœa and anxious countenance, with bulging and want of respiratory movement of the affected side and displacement of the heart, may be noticed, as in pneumothorax or pleurisy with effusion. If the opening into the pleural cavity be valvular, so that air enters more readily than it escapes, the signs are more marked. This may occur at different times in the course of the disease.

Palpation.—The vocal fremitus is diminished or absent on the affected side. The heart may be felt out of its normal position.

Percussion.—Above the level of the fluid the note is tympanitic, as in pneumothorax. Over the fluid there is marked dullness or flatness, as in pleurisy with effusion or hydrothorax. Not infrequently about the union of the two there is amphoric (jug, metallic) resonance on percussion. The line of dullness or flatness and tympanicity, changes markedly with position of the patient. In these cases, also, the upper end of the fluid is horizontal, instead of being slightly curved as in pleurisy with effusion.

Auscultation.—The respiratory murmur is weakened or absent, or else it is amphoric (jug, metallic). The

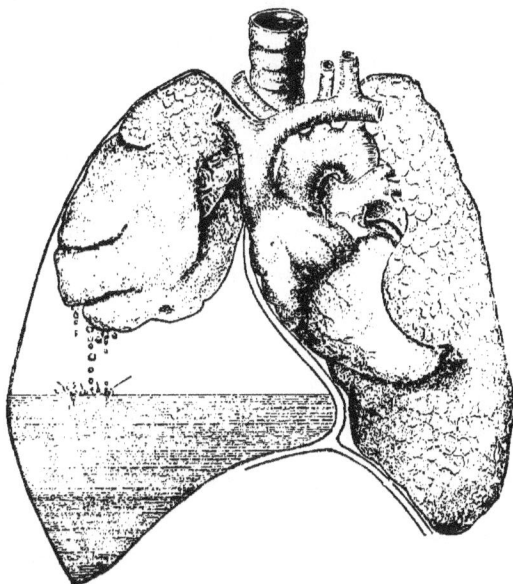

FIG. 17.—Diagram of Pneumohydro-thorax, showing fluid and air in right pleural cavity, right lung compressed and displacement of organs to the opposite side. Also the supposed dropping of fluid.

vocal resonance is also weakened, or sometimes amphorophony, or even amphoriloquy, may be present.

Succussion.—This is the act of shaking the patient while auscultating at the same time. On succussion the splashing sound of the fluid in the cavity is heard, often by the patient as well as the physician.

Metallic (amphoric, jug) tinkle is often heard in this disease, and by many it is supposed to be produced by the dropping of fluid from above into the fluid below. It may be possible that in such a large cavity as is usually represented, fluid may have sufficient distance to fall to produce the sound; but even here it is more probably due to the bursting of bubbles formed by inspiration, or by shaking the patient. The distinctive sign of the disease, and one which prevents its being mistaken for any other, is the splashing sound of the fluid in the cavity heard on succussion. It is impossible to obtain this sound in simple pleurisy with effusion, since there is no air in the cavity, and hence the fluid there cannot be shaken any more than it could be in a bottle, or other vessel, filled up to the cork. In all suspected cases, succussion should be tried.

CHAPTER V.

Summary of adventitious sounds, and of the changes in the normal respiratory murmur and vocal resonance produced by disease.

ADVENTITIOUS SOUNDS, as previously stated, are wholly new and abnormal sounds produced by disease, and are not modified normal sounds. They have been variously divided and arranged by different authors, but for simplicity as well as convenience they may be classified into (1) râles or rhonchi, (2) friction sounds, and (3) splashing sounds.

I. RÂLES.

Râles, or rhonchi (rattles), have also been grouped differently by different authors, but they may be reduced to three varieties: (1) dry, (2) moist, and (3) indeterminate. Each of these varieties will now be considered separately.

1. DRY RÂLES.

There may be many varieties of dry râles, but two are sufficient to include them all: (1) the sonorous and (2) the sibilant.

Sonorous Râles are loud, low-pitched, dry râles, made on inspiration, expiration, or both. They may be produced in the larynx, trachea, or larger bronchi. In the larynx they may be caused by spasm of the glottis, as observed in laryngismus stridulus, croup, and

whooping-cough, growths within the larynx, and pressure on the recurrent laryngeal nerve from aneurism or
other tumor. In the trachea they may be due to pressure without from some tumor, aneurism for example,
or growths within, as polypi, or œdema, inflammatory
exudation, or constriction due to old cicatrices and
the like. Such râles are termed stridor, and the breathing is said to be stridulous. If the pressure, or obstruction, or constriction, be sufficiently marked, sibilant as well as sonorous râles may be produced in these
localities. These râles made in the larynx or trachea
are, however, conveyed all over the chest usually, and
should be differentiated by means of the stethoscope;
an easy matter, as they are much louder at the site of
their production than elsewhere Sonorous râles are
made in the larger bronchi from narrowing of their
calibre by external pressure from some tumor, stricture due to old inflammation, spasm of the muscular
coats, or tumefaction of the lining mucous membrane,
or else the vibration of viscid mucus within the tubes.
These râles are generally transient, and change on
coughing. For this reason, it would appear that they
are more frequently due to varying spasm or vibrating
mucus that is removed by coughing. Where the cause
is permanent, the râles are few in number, and often
changeable, though less so, showing that even in these
cases, vibrating mucus, from localized irritation, is often
an important factor. Both sonorous and sibilant râles
are heard, especially in the early or dry stage of bronchitis and in spasmodic asthma. In the latter disease
they are also sometimes said to be mewing and chirping.

Sibilant râles are high pitched, whistling, dry râles,

made on inspiration, expiration, or both. As already remarked, they may be made in the larynx or trachea, if the calibre be diminished sufficiently. For the same reason they may also be made in the larger bronchi. Usually, however, they are made in the smaller bronchi. The causes of their production are the same as for the sonorous or snoring râles, and, like them, are heard es-

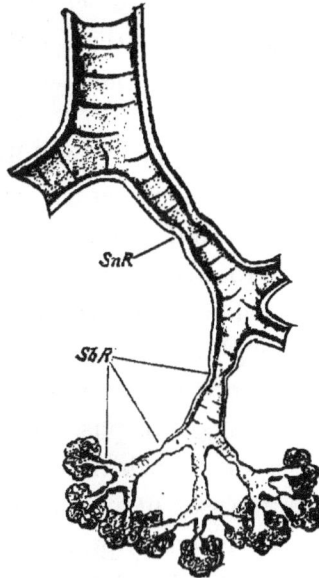

FIG. 18.—Sonorous and Sibilant Râles. *SnR*, Sonorous Râles; *SbR*, Sibilant Râles.

pecially in the dry stage of bronchitis, and in an attack of asthma. They are also changeable, being heard more distinctly now in one place and then in another.

2. MOIST RÂLES.

These may occur in the larynx, trachea, bronchi, air-cells, or in the pleural cavities. Laryngeal and tracheal râles may also be moist as well as dry. Moist tracheal râles occurring just before death, as they often do, are

11

commonly called death rattles. These râles may also be heard over the chest, and their locality should be differentiated by means of the stethoscope. In cases where râles of any kind are heard over the chest, the patient should be directed to clear the throat, so as to get rid, if possible, of any laryngeal or tracheal râles that might exist.

Many varieties of moist râles occurring in the bronchi are described by various authors, but three are sufficient to include them all: (1) mucous, (2) submucous, and (3) subcrepitant.

Mucous râles are large, moist, bubbling râles made in the larger bronchi, and are heard both on inspiration and expiration, since the tidal air acts on the fluid in the tubes both on entering and leaving them. These râles are usually attended with expectoration, and they change about on coughing. Fluid of any sort in the larger bronchial tubes will give rise to mucous râles, whether it be mucus, blood, or pus. We therefore find them in such diseases as bronchitis, unless the secretion be very scant, certain cases of pulmonary hemorrhage, abscess, and pulmonary consumption. In general bronchitis the râles are bilateral, but usually few in number. If not heard at all on first listening, they may be developed by coughing. In other diseases where the bronchitis would be local, the râles would also be localized.

Submucous râles are moist, bubbling râles, rather smaller than the mucous, and are made in the medium-sized bronchi. They are also heard both on inspiration and expiration, are attended with expectoration, and are changed by coughing. They are produced in the same way, and are due to the same causes, as the mucous râles (see fig. 8, p. 38).

Subcrepitant râles, or muco-crepitant râles, are the finest moist bronchial râles, and are made in the finer (ultimate) bronchial tubes, chiefly on inspiration, and are not as easily changed as the mucous and submucous by coughing. This is accounted for by the fact that inspiration is a greater force than expiration, as affecting the finer tubes and air-cells. Hence the mucus in these localities is overcome with greater force on inspiration than expiration, and the subcrepitant râles are consequently produced on inspiration rather than expiration. They may be, and often are, entirely wanting during expiration. Instead of being caused by the bursting of very fine bubbles, these râles are sometimes caused by the forcible separation of agglutinated tube walls, and then they are invariably heard only on inspiration. In capillary bronchitis they are heard on both sides, usually posteriorly and low down. In pulmonary œdema they are also sometimes heard as the serous fluid enters the tubes (see Œdema). The râle also occurs in the third stage of lobar pneumonia, due to secondary local bronchitis and liquefying exudation. In pulmonary hemorrhage from any cause these râles are produced if blood enters the finer tubes, and hence are often heard in hemorrhagic infarction. Pus in these finer tubes also gives rise to it, whether the pus be due to purulent bronchitis, rupture of an abscess, or perforation from empyema. This râle is also the adventitious sound usually heard in the first stage of phthisis, although it may evidently occur in any stage.

Crepitant râles are made in the air-cells, and are the

only vesicular râles that exist. They are very fine, uniform, crackling râles, heard only at the (tip) end of inspiration, and are unchanged by coughing. They are caused by the forcible separation of agglutinated cell walls, as in the first stage of lobar pneumonia, or agitation of thin fluid in the air-cells, as in pulmonary œdema. They are heard also in the third or resolving stage of lobar pneumonia, where it is known as the râle redux, or râle that has returned, or come back. The crepitant râle, then, and not the subcrepitant, as is so often erroneously stated, is the râle redux, or returned râle. The subcrepitant râle is not a part of the physical signs of any stage of pneumonia but the third, in which it has not come back, but is heard for the first time. The crepitant râle, however, which was heard in the first stage, and lost in the second after the air-cells had become obliterated, now comes back, or is redux, in the third stage. The crepitant râles heard in pulmonary œdema are louder and more liquid in quality than those of pneumonia. The crepitant râle may also be heard in pulmonary consumption in any stage, if the conditions for its production exist, and other adventitious sounds, usually present, are not so loud as to obscure it.

Mucous Click.—A single fine, high-pitched moist click that is usually unchanged by cough, may be heard at or near the end of inspiration over incomplete consolidation in the first stage of phthisis. As Loomis says, it sounds like an isolated subcrepitant râle. When it is really of mucous origin, it is no doubt due to the sudden passage of air through a fine bronchial tube obstructed by pressure without, due to tubercle, for in-

stance, or viscid mucus within. It is a very important
adventitious sound in the commencement of pulmonary consumption. But may it not sometimes also be
of pleuritic origin?

Gurgles are moist, bubbling râles made in a cavity,
and are large or small, or low and high pitched, according to the size of the cavity and amount of consolidated
tissue intervening. They are made either during inspiration or expiration, but inasmuch as the air usually enters a cavity with more force than it leaves, gurgles are consequently apt to be louder on inspiration
than expiration. Sometimes they are heard only on
inspiration. Moreover, during inspiration, the direction of the current is toward the ear of the listener, instead of from it, as in expiration. If the cavity be full
of fluid there may be no gurgles. None will be heard,
also, if the cavity be empty, or if the opening into it be
such as to prevent fluid from being agitated by the air
entering or leaving it. If the opening into the cavity
becomes stopped with a plug of viscid mucus, a clot
of blood, or other material, gurgles which were heard
before now immediately cease. On coughing and removing the obstruction they at once return. (See colored plate showing cavity.)

Metallic Tinkle, or Amphoric Tinkle.—If the cavity,
whether it be pulmonary or intra-pleural, have hard,
smooth walls, and be of sufficient size to act as a resonant chamber, in other words, if it be an amphoric cavity, and if it contain a small amount of viscid fluid
with the tube opening under it or into it, so as to produce explosion of bubbles, metallic, or amphoric tinkle
is apt to result. Or it may be produced by the vibra-

tion of viscid mucus. It is usually heard most distinctly on inspiration, like ordinary gurgles, and is not
constant. It often disappears on coughing, or it may
be developed by coughing. According to Walshe, such
adventitious sound need not be situated in the cavity
at all, if it be near enough so that the sound shall be
echoed in the cavity. Metallic tinkle is also sometimes
produced by speaking or coughing. It is probably
never due to dropping of fluid in a pulmonary cavity,
since there could hardly be distance enough for it to
fall with sound sufficiently distinct to be heard. Even
in pneumo-hydrothorax, metallic tinkle may be regarded as a musical râle of amphoric quality, produced
by the bursting of bubbles rather than the dropping
of fluid. For that reason it has been classified with
the moist râles.

Intra-pleural Moist Râles.—Pleuritic friction sounds
are easily distinguished from other adventitious sounds;
but occasionally we hear sounds, evidently of intra-
pleural origin, that exactly imitate moist bronchial or
vesicular râles. Some authors go so far as to say that
the crepitant, as well as the subcrepitant râle, is always intra-pleural. In the same way other intra-
pleural adventitious sounds, resembling mucous and
submucous râles, are sometimes heard, so that, as Da
Costa says, no human ear can tell the difference by the
quality alone. How are we to distinguish between
them? This subject has already been alluded to (see
Bronchitis). If the râles are localized, unilateral,
unchanged by coughing, peripheral and superficial,
and unattended with expectoration, they are usually
intra-pleural. If, on the other hand, they are bilateral,

changed by cough, attended with more or less expectoration, deep-seated, and generally distributed over the
chest, they are almost surely bronchial. But how can
we distinguish the crepitant râle made in the air-cells
from the intra-pleural crepitation? Sometimes it is
difficult or even impossible. This, however, does not
necessarily make the two identical any more, for instance, than it makes thoracic aneurism and pleurisy
with effusion identical, because one has been mistaken
for the other until *post-mortem* examination revealed
the true state of the case. In pulmonary œdema we
have the crepitant râle bilateral and low down posteriorly. There is also watery expectoration and the
cause of œdema (see Œdema). Intra-pleural crepitation
is rarely bilateral, and then is unattended with expectoration, unless there be complication. In the first and
third stages of lobar pneumonia it is more difficult to
say that the râle is not intra-pleural. Even in these
cases the râle sometimes follows the outline of the
lobe too closely to say that it is always intra-pleural.

Intra-pleural moist râles do not require actual inflammation of the pleuræ for their production. Perverted nutrition of the membrane from any cause may
give rise to a viscid, glutinous secretion, instead of the
normal lubricating material, so that the pleuræ, instead of gliding noiselessly on each other during respiration, will produce sounds which, as already stated,
may be, and often are, identical in quality with vesicular, bronchial, tracheal, or laryngeal moist râles.

3. Indeterminate Rãles.

Indeterminate rãles include all other rãles not embraced in the foregoing classes and varieties. They are crackling and crumpling sounds, partly moist and partly dry, produced on inspiration or expiration, or both, and it is impossible to determine with certainty whether they are of intra-pleural, pulmonary, or bronchial origin. Flint states that they are found usually early in phthisis. They may be heard during any stage, but particularly, perhaps, after the disease has advanced sufficiently to give rise to broken-down tissue and complicated pathological conditions. The following table in regard to rãles may be of use:

Rãles or Rhonchi. 3 classes.	1. Dry Rãles.	Stridor—produced in the Larynx. Sonorous—produced in the Large Bronchi. Sibilant—produced in the Small Bronchi.
	2. Moist Rãles.	Laryngeal—produced in theLarynx. Tracheal—produced in the Trachea. Mucous—produced in the Large Bronchi. Submucous—produced in the Medium Sized Bronchi. Subcrepitant—produced in the Small Bronchi. Crepitant—produced in the Air-cells or vesicles. Mucous Click—produced in a Small Bronchus. Gurgles—produced in Cavities (pulmonary). Metallic (amphoric) Tinkle—produced in Cavities Interpleural moist rãles which may simulate any of the above moist rãles.
	3. Indeterminate Rãles.	Partly moist, partly dry, crackling and crumpling sounds, whose exact origin and mode of production are unknown.

II. Friction Sounds.

Friction sounds are due to pleuritic inflammation. Their true pathological significance was first pointed out about 1819 by Honoré, of Paris, a contemporary of Laennec. The membrane, which in health was moistened with a lubricating secretion, so that the two opposite layers glided noiselessly upon each other during respiration, now becomes dry and rough, or agglutinated. For this reason, grazing, rubbing, rasping, grating, rumbling, or creaking sounds may be produced. The sounds are heard chiefly on inspiration, and are usually interrupted, so that several may occur during one inspiration. They are also heard sometimes during expiration. As already stated, pleuritic friction sounds are heard more distinctly usually on inspiration than expiration (p. 140). As Flint correctly states, these sounds are usually more distinctly heard after removal of the fluid, if any exist, than before, as the surfaces of the pleuræ are by that time more roughened than at first. These friction sounds are of variable duration, depending upon the pathological condition. When well marked, especially in those having thin chest walls, friction fremitus may be felt on palpation.

III. Splashing Sounds.

Succussion is the act by which splashing sounds are produced. It consists in shaking the patient while the ear is placed to the chest. By this means the splashing of fluid in a cavity containing air and fluid is heard, as in pneumo-hydrothorax (or pneumo-pyothorax). It might be possible, also, in a large pulmonary cavity

containing air and fluid. It is usually, however,
pathognomonic of pneumo-hydro (or pyo) thorax.

CHANGES IN THE RESPIRATORY MURMUR.

I. CHANGES OF INTENSITY.

The sound of the respiratory murmur, as well as of
the voice, arrives at the chest walls in health, first,
through convection along the tubes, and secondly, refraction (diffusion) in the air-cells. But for this refracting power of the healthy lungs, due to the presence of
the air-cells, the intensity of the respiratory murmur
and of the voice sounds would be greatly increased, as
occurs in solidification. On the other hand, the sounds
would be diminished or absent, according to the amount
of obstruction in the tubes to convection, increased refraction, as in emphysema, and interruption, as in pleuritic thickening or effusion.

We see, then, that these, as well as other sounds on
their way to the chest walls, may, according to the conditions present, be subjected to (1) convection, (2) obstruction, (3) refraction (diffusion), (4) conduction, and
(5) interruption. On the chest walls, or along strings
of adhesion, they may be transmitted, or extended.

In health, then, when convection along the tubes is
perfect, and there is only normal refraction (diffusion)
in the air-cells, we hear the normal respiratory murmur,
which will be laryngeal, tracheal, bronchial, vesiculo-
bronchial, or vesicular, according to the locality.

The intensity of this murmur will be weakened or
suppressed (also termed diminished or absent); (1) in
proportion to obstruction to convection in the tubes,
either from growths, mucus, pus, blood, or other ob-

stacles within, or stricture, from old inflammation or compression of the tubes from aneurism, cancer, hydatids, or some other cause; (2) increased refracting (diffusing) power in the lungs, due to dilatation of the air-cells, as in chronic general emphysema; or (3) interruption from pleuritic thickening or effusion. Secondarily, it would be modified also by interference with the proper expansion of the lungs and chest walls from pain, pleuritic adhesions, and deformities.

The intensity may be slightly increased (exaggerated) in two ways. First, without any change in quality or rhythm, as when one lung or part of a lung is temporarily *vicariously* emphysematous from doing extra work, owing to crippling of the other lung or part of a lung. It is simply louder than normal, and is sometimes termed hyper-vesicular or supplementary. It differs somewhat from puerile breathing, which is heard in children under the age of puberty, whose lungs are not developed in proportion to their bronchial tubes. Both are loud (rude), but the puerile respiratory murmur has more of a bronchial element in it. Neither are necessarily harsh (rough), this quality depending on the roughness of the mucous membrane lining the larger bronchial tubes, and caused by the friction of the tidal air

Secondly, it may be slightly increased (exaggerated), with change in the quality and rhythm, as seen in the vesiculo-bronchial breathing, due to incomplete consolidation with corresponding increase of conduction, and diminution of refraction, in the lungs, owing to obliteration, in part, of some air-cells; as seen in the first stage of phthisis pulmonalis.

The intensity of the murmur will be markedly increased if refraction is replaced by conduction; in other words, if the air-cells are replaced by solidified lung tissue, as occurs, for instance, in the second stage of lobar pneumonia, which offers a homogeneous medium for the conduction of sound.

II. RHYTHM.

Prolonged expiration and divided respiration are the two principal changes in the rhythm. In the normal vesicular respiratory murmur, inspiration is about four times longer than expiration, and the two are continuous. Moreover, expiration is lower in pitch than inspiration. Now, in a paroxysm of asthma, and in chronic emphysema, the expiration is prolonged, but is not usually changed in pitch or quality. In a paroxysm of asthma the expiration is not only prolonged, but inspiration is much shortened by being deferred, that is to say, it is not heard in the commencement. The rhythm in such a paroxysm is just the reverse of what it is in health, that is to say, expiration is four times longer than inspiration, or even longer. In emphysema, expiration is also prolonged and inspiration deferred, but it is never observed that expiration is four times longer than inspiration, unless there be also marked obstruction to the exit of air from the presence of mucus, or other cause. In bronchitis, also, expiration is prolonged in proportion to the obstruction from mucus, especially in capillary bronchitis; but the respiratory murmur is not otherwise changed, unless in some cases it may become harsh or rough (see Changes in Quality). In asthma and bronchitis, the expiration is prolonged, because of obstruction to the egress of air, either from spasm of the muscular coats of the tubes, or mucus within the tubes, or tumefaction of the bronchial mucous membrane. In general emphysema, on the other hand, expiration is prolonged on account of the rigid dilatation of the thorax, caused by loss of resiliency of lung tissue, rigidity of the costal cartilages, and crippling of the diaphragm, leaving the muscular coats of the bronchial tubes to perform the act. If obstruction in the tubes, due to co-existing bronchitis, also is well marked, the expiration will be still more prolonged, but usually the expiration of emphysema is not so much prolonged as to be four times the length of inspiration, as in a paroxysm of asthma, where obstruction is marked on account of spasm as well as mucus and tumefaction. The reason for prolonged expiration in such cases has already been fully explained (see Asthma).

In consolidation of lung tissue, expiration is also prolonged, but the murmur in these cases differs from those just mentioned by having the pitch always raised, on account of its passing from one medium to a denser medium, with shorter vibrations and the quality changed. Instead of being blowing in quality, it is more or less tubular. The prolongation is not due so much to the crippling of the expiratory forces, or to obstruction to the egress of air, as it is to the fact that the incompletely or completely solidified lung tissue, being a better conducting medium, enables one to hear the murmur more distinctly and for a greater length of time.

The respiratory murmur may be divided—that is,

the inspiration and expiration may not be continuous. This is usually, if not always, due to consolidation to a certain degree. In case of incomplete consolidation, inspiration and expiration are not continuous, by inspiration being unfinished. The gap between the two will be marked in proportion to the extent and completeness of consolidated tissue. When a whole lobe is consolidated, as in the second stage of lobar pneumonia, for instance, the break between inspiration and expiration is well marked.

Wavy, jerky, or cog-wheeled respiratory murmur is sometimes heard in the first stage of phthisis. It is usually attributed to the sudden overcoming of stricture or obstruction in a bronchial tube, from pressure of tubercle, or mucus within. This, however, is doubtful, since it is often heard in perfectly healthy chests of those who are nervous, hysterical, or have palpitation. It is more often due to palpitation of the heart than anything else, and is oftener heard in women than men. Taken by itself as a physical sign of disease of the lungs, it is worthless. It is usually heard on inspiration, but may also be heard on expiration, or both.

III. QUALITY.

Besides being changed in intensity and rhythm, the respiratory murmur may also be changed in quality. Instead of its being purely vesicular in quality, as heard over the left subclavicular region, it may become vesiculo-bronchial. The late Dr. Austin Flint termed it broncho-vesicular, but, as Da Costa suggests, the order of occurrence, as actually heard, being first vesicular and then bronchial, it is more correct to call it vesiculo-bronchial. It was formerly, and is now by some, termed rude (loud) respiration. Others, again, call it harsh or rough. Neither of these terms is descriptive of the true condition, nor is the murmur necessarily harsh, rough, or even rude. A respiratory murmur is harsh or rough if the mucous membrane of the larger tubes is in a harsh or rough condition, so that the tidal air, by friction against it, in passing in and out, has the quality of harshness or roughness imparted to it. For this reason, any respiratory murmur, excluding the purely normal vesicular, may be harsh or rough.

In vesiculo-bronchial breathing, expiration is prolonged, raised in pitch, and more or less tubular in quality, in proportion to obliteration of air-cells, by which the vesicular element is lessened, as in incomplete consolidation from some cause, or proximity to bronchial tubes. For the latter cause, we obtain a normal vesiculo-bronchial breathing in the right subclavicular region of a healthy chest. Vesiculo-bronchial breathing is also among the early signs of phthisis, while consolidation is incomplete.

Bronchial breathing is significant of complete consolidation of lung tissue, as in the second stage of lobar pneumonia. Here all vesicular quality is lost, and both inspiration and expiration become tubular in quality. Expiration is as long as inspiration, or longer; the pitch of both is raised, expiration being usually higher than inspiration, and the intensity is greatly increased in the sense, not of volume, but concentrated amount. It sounds somewhat like blowing across the mouth of the stethoscope. It may be imitated by putting a piece of liver in a tin, or other tube, covering

both ends with a thin membrane, and then listening
through it to tracheal respiration. Over a cavity we
usually hear cavernous breathing, which is blowing in
quality, giving the idea, as Flint states, of air passing
in and out of a hollow space. It differs from bronchial
breathing by being usually low pitched and blowing in
quality, whereas bronchial breathing is necessarily
high pitched, the sound of the murmur having passed
from one medium to a denser medium, and tubular in
quality. Both are more intense than the normal respiratory murmur, but cavernous breathing is more intense in the sense of volume, bronchial breathing in
concentrated amount.

When consolidated lung tissue is extensive near a
cavity, there is sometimes heard a mixture of cavernous and bronchial qualities, or the caverno-bronchial
breathing. Flint was the first to describe this kind of
respiratory murmur, and termed it broncho-cavernous.
In nearly all the cases cavernous quality comes first,
and is followed by the bronchial, though there is no
reason why this should always be so. For this reason,
however, it is termed caverno-bronchial, rather than
broncho-cavernous.

Should the cavity be large, situated near the apex,
and have hard, smooth walls, a free opening communicating with a bronchial tube, and not contain much
fluid, amphoric respiration, or jug-breathing, may be
heard. It sounds like blowing into the mouth of an
empty jug (amphora) or bottle. The pitch of this
breath sound will vary with the size of the cavity, its
mouth and other conditions, but it is by its quality,
not pitch, that it is distinguished.

CHANGES IN THE PECTOROPHONY OR VOCAL RESO-
NANCE.

The sounds of the patient's voice, as heard over the larynx, trachea, or any part of the chest walls by the auscultator, may be divided into two classes: (1) that in which only the voice is heard, and (2) when speech or articulation is heard. To the first class the termination ophony (*phonos*, voice) is applied, to the latter, iloquy (speech). The sound of the voice, therefore, as normally heard over the larynx, is termed normal laryngophony, and over the trachea, normal tracheophony or trachophony. As the speech is oftener heard in these localities than the voice simply, so do we more frequently hear normal laryngiloquy and normal trachiloquy over these organs.

I. PECTOROPHONY.

Pectorophony, or chest voice, is the sound of the voice, or the vocal resonance, heard over the chest, without our being able to distinguish the articulate words as spoken by the patient.

Normal pectorophony, or normal vocal resonance, as it is more commonly called, when heard over pulmonary vesicular tissue, is a distant, diffused, indistinct, buzzing sound, with a somewhat low pitch, corresponding to the pitch of the patient's voice. A low-pitched, loud, harsh voice, other things equal, yields more intense pectorophony than a high-pitched, weak voice, in the sense of volume. For this reason, men usually have normally more intense pectorophony than women, and grown people of both sexes than

12

children. Thin chest walls, other things equal, also
favor the production of pectorophony.

Diminished pectorophony, or weakened vocal resonance, occurs in those cases where there is obstruction
in the bronchi to the convection (conveyance) of the
voice sound, as in bronchitis with abundant mucous
secretion, pus or blood, polypi, stricture of the bronchi, from old inflammation, or their compression from
some cause, as cancer, aneurism, hydatids, and other
tumors. Also when there is increased refractive power
of the lungs, as in chronic general emphysema, with
permanent dilatation of the air-cells, or when there is
interruption to the transmission of the voice from
pleuritic thickening or effusion. In some of these
cases, whispering pectorophony may be entirely absent
or suppressed. In general, the same conditions for
weakening or suppressing the respiratory murmur
apply to the whispered voice.

Exaggerated pectorophony, or exaggerated vocal
resonance, occurs when the intensity is increased from
any cause. It is heard normally in the right sub-
clavicular region, on account of proximity to the right
bronchial tube, as already stated. But it is also heard
over incomplete solidification of lung tissue, owing to
the better conducting power of the latter. It is therefore one of the early signs of incipient phthisis. Under
these conditions, the sound of the voice is nearer, less
diffused, and more distinct and intense, in the sense.
of concentrated amount.

Bronchial pectorophony, bronchophony, or bronchial voice, is heard when all vesicular quality is lost,
owing to complete solidification of lung tissue. The

voice sound comes to the ear of the auscultator directly from the bronchial tubes, through the solidified lung tissue. It is near, concentrated, and distinct, or the very opposite of normal pectorophony (vocal resonance). The pitch of bronchophony is necessarily high, owing to the transition of the sound from one medium to a denser medium.

In some instances, bronchophony, instead of being near and strong, sounds as if it were distant and weak. Weak or distant bronchophony is caused by the intervention usually of pleuritic thickening or effusion. Obstruction to convection in the bronchi from compression, or stricture, or a plug of viscid mucus, might cause bronchophony to be weakened or · even suppressed.

Ægophony, so named from its resemblance to the bleating of a goat, is bronchophony made tremulous by vibrating fluid in the pleural cavity usually, though it may occur in some rare cases of pulmonary cavities and other disease. It is more of a clinical curiosity now than it is of any real value, since, in cases of doubt and necessity, the aspirating needle may be used to determine the presence of fluid. Besides being tremulous, ægophony is more or less weakened by the pleuritic conditions present. It also usually possesses a nasal quality for some reason.

Cavernous voice, or cavernophony, is heard when listening over a cavity. It differs from bronchophony, for while the latter is high-pitched and tubular in quality, with an increased intensity in the sense of concentrated amount, cavernophony is often low pitched, sepulchral in quality, and has its intensity increased in the sense of volume.

Amphoric voice, or amphorophony (jug voice), is simply cavernophony possessing the quality of amphoricity, and resembles the sound of the voice in an empty jug or other vessel. It has a peculiar ringing, metallic, or amphoric quality, and may be attended, like the breathing, or the cough, by amphoric, or metallic echo. The terms amphoric and metallic are used to convey the same idea, the former referring to the particular kind of cavity, the latter to the material, which imparts the quality to sounds. Should the patient whisper instead of speaking out loud, as was first recommended by the late Dr. Austin Flint, we would have some variety of whispering pectorophony. In some cases, it might be absent or suppressed; in others, exaggerated, and then Flint termed it exaggerated bronchial whisper. In the same way we would have whispering bronchophony, cavernous whisper or whispering cavernophony, and amphoric whisper or whispering amphorophony. The following table gives a summary of the voice sounds:

Laryngophony
Tracheophony

Pectorophony
{
Normal Pectorophony, or vocal resonance.
Diminished or weak (the whisper may be absent or suppressed).
Exaggerated (slightly increased).
Bronchophony—bronchial voice (may be weak).
Ægophony—goat's voice (tremulous).
Cavernophony—cavernous voice.
Amphorophony—amphoric (jug) voice.
}

II. PECTORILOQUY.

Pectoriloquy, or chest speech, is the speech (articulate words) of the patient as heard through the chest walls by the auscultator. The term was first used by Laennec, of Paris, about 1820, and always had reference to cavities, since it was thought that only over cavities could the articulate words of the patient be heard. It is, however, quite evident that pectoriloquy may be obtained in other conditions than cavities. It is sometimes heard over perfectly healthy chests, especially among those who have thin chest walls and a loud respiratory murmur, as among some women and children. It is quite often heard over the larynx and trachea, normally.

Bronchial Pectoriloquy, or Bronchiloquy.—Instead of bronchophony simply, we sometimes also hear the articulate words of the patient over consolidated lung tissue. Indeed, Guttmann states that there is no difference between bronchophony and pectoriloquy. But as bronchophony differs from cavernophony, both being pectorophony, so does bronchial pectoriloquy, or bronchiloquy, differ from cavernous pectoriloquy or caverniloquy, the former being necessarily high-pitched, the latter usually low-pitched. In case of whispered voice, the former is also tubular in quality, the latter blowing. The intensity of the former is also increased in concentrated amount, the latter in volume.

Cavernous pectoriloquy, or caverniloquy, is the speech of the patient as heard over an ordinary cavity. As Flint truly remarks, one must not expect to be able to carry on a conversation with his patient through a cavity, but if some syllables of some simple phrase or words, as one, two, three, be heard, it is sufficient to establish pectoriloquy of any variety.

Amphoric pectoriloquy, or amphoriloquy, is the

speech of the patient with an amphoric or metallic in-
tonation, as if speaking in the mouth of an empty jug.
It is significant of amphoric cavity. If the patient
whispers instead of speaking out loud, we then hear
whispering laryngiloquy, trachiloquy, or pectoriloquy,
of whatever variety the latter may be. The following
table gives a summary of varieties of speech (articulate
voice) sounds:

Laryngiloquy.
Trachiloquy.

Pectoriloquy
{
  Normal Pectoriloquy.
  Bronchiloquy—Bronchial speech (may be weak
    or tremulous).
  Caverniloquy—Cavernous speech.
  Amphoriloquy—Amphoric speech.
}

CHAPTER VI.

THE HEART.

THE HEART is a hollow organ of striated or voluntary muscular tissue, but so presided over by the sympathetic nervous system that its movements are, with very rare exceptions, wholly involuntary. Only in extremely rare cases has the individual been able to cause the heart to beat fast or slow at will. The fact that it is of the striated or voluntary muscular tissue is of great importance in connection with certain dynamic cardiac murmurs, to be described hereafter. Normally, the heart is conical in shape, and, in the adult, five inches long, three and a half inches broad and two and a half inches thick. It weighs from ten to twelve ounces in men, and in women from eight to ten ounces.

The heart is obliquely situated within the thorax, between the lungs, and is inclosed by the pericardium. The base, directed upward and backward to the right, is on a level with the upper borders of the third costal cartilages, extending half an inch to the right, and one inch to the left of the sternum; the apex, forward and downward to the left, corresponds to a point between the fifth and sixth costal cartilages (fifth intercostal space), two inches below and one inch within the left nipple, according to Gray. According to others, it is an inch and a half below, and half an inch within, the left nipple, and this is probably more nearly correct.

According to Flint, the apex of the heart in health should fall a little within the mammillary line. Of course these rules apply only to those cases where the nipple is in its normal position, for sometimes it is displaced by deformity, or large size of the gland, as in nursing women or those who have borne children. The force of the normal apex-beat differs in different cases. In some it is perceptible on inspection, in others not, and in some cases it may not even be felt on palpation. This is accounted for chiefly by difference in the thickness of the chest walls and size of the ribs. Among those having thick chest walls with wide ribs, the impulse of the heart will not usually be so perceptible as among those who have thin chest walls and narrow ribs with correspondingly wide intercostal spaces. The normal heart beats more forcibly in some persons than others, and the impulse also differs somewhat with position of the body.

Outline of the Heart.—The base corresponds to a line drawn across the sternum along the upper borders of the third costal cartilages, extending half an inch to the right and one inch to the left of the sternum. A line so drawn is termed the base line of the heart. The left border corresponds to a line curving outward, but within the left nipple, from the left end of the base line, down to the apex. This border is formed by the left ventricle. The right border of the heart consists of a right border proper, formed partly by the right auricle, and partly by the right ventricle, and a lower border formed chiefly by the right ventricle. Draw a line from the apex horizontally to the median line of the sternum, to correspond with the lower border, thence curving upward and slightly outward to the right end of the base line, to form the right border proper.

Areas of Cardiac Dullness.—Auscultatory percussion (p. 33) is the best method for accurately mapping out the limits of the heart, the patient being in the erect position, unless one has no assistant, and then the recumbent position is best. There are two areas of dullness, the deep and superficial. The whole area of dullness, including the deep and superficial, extends vertically from the upper borders of the third costal cartilages to the upper border of the sixth, and transversely from a point a little within the left nipple to about half an inch to the right of the sternum.

The deep area of dullness corresponds to that portion of the heart covered up by lung tissue, and is increased with enlargement of the heart from any cause.

The superficial area of dullness is somewhat triangular in shape, with little lung tissue over it. This area is bounded below by a line drawn horizontally from the apex to the median line of the sternum; on the right, by the median line of the sternum up to the level of the upper borders of the fourth costal cartilages; and on the left, by a line drawn from the last-named point to the apex. This last line curves outward, but falls within the left nipple. The superficial area of dullness is diminished at the end of a full inspiration and in emphysema; it is increased by ventricular enlargements and pericardial effusion. It is formed by the right ventricle, except at the apex, which is composed of the left ventricle. This latter fact is very important in connection with murmurs made within the left ventricle.

The Circulation.—The heart is divided lengthwise into right and left, or venous, and arterial halves, and these two are divided crosswise, so as to form two upper and two lower compartments—four in all—the two upper compartments being the right and left auricles respectively, and the two lower being the right

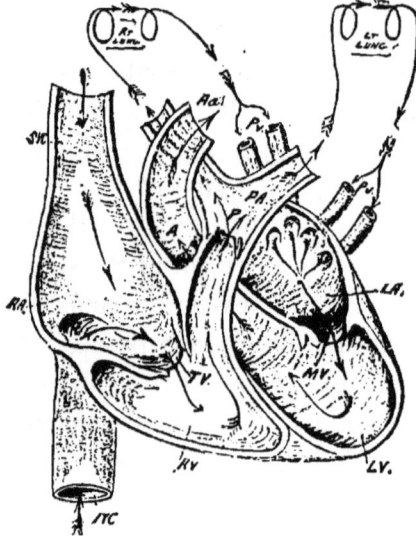

FIG. 19.—Normal Blood Currents in the Heart and Relative Position of the Ventricles, Auricles and Great Vessels. *IVC*, Inf. Vena Cava; *SVC*, Sup. Vena Cava; *RA*, Rt. Auricle; *TV*, Tricuspid Valves; *RV*, Rt. Ventricle; *P*, Pulmonary Valves; *PA*, Pulmonary Artery; *Pv*, Pulmonary Veins; *LA*, Left Auricle; *MV*, Mitral Valves; *LV*, Left Ventricle; *A*, Aortic Valves; *Aa*, Arch of Aorta.

and left ventricles. The right (or venous) side of the heart is situated in front of, slightly above and to the right of the left side, so that in looking at the normally situated heart, from the front, we see the right ventricle and right auricle, and only a small part of the left auricle, and a narrow strip of the left ventricle, which extends down further than the right ventricle, and forms the apex.

PLATE VI.

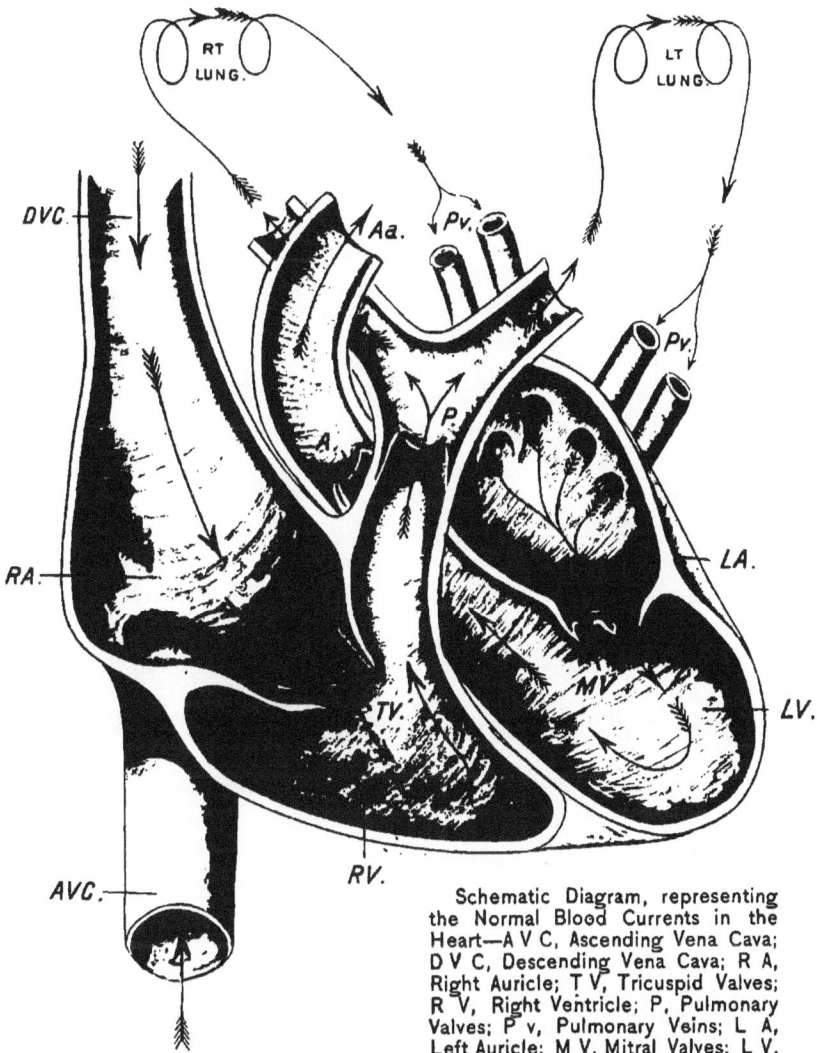

Schematic Diagram, representing the Normal Blood Currents in the Heart—A V C, Ascending Vena Cava; D V C, Descending Vena Cava; R A, Right Auricle; T V, Tricuspid Valves; R V, Right Ventricle; P, Pulmonary Valves; P v, Pulmonary Veins; L A, Left Auricle; M V, Mitral Valves; L V, Left Ventricle.

A, Aortic Valves; A a, Arch of Aorta.

The pulmonary artery is about two inches long, and arises from the left side of the base of the right ventricle, in front of the aorta, at a point corresponding to the junction of the left third costal cartilage with the sternum. The left auricle lies deeply behind it. It ascends obliquely upward and outward across the second left intercostal space near the sternum, and divides under the arch of the aorta, behind the second left costal cartilage, into a right and left branch, one for each lung. The second left intercostal space is also called, therefore, the pulmonary (pulmonic) interspace. The pulmonary artery carries venous blood from the right ventricle into the lungs.

The aorta arises from the upper part of the left ventricle behind, and a little below, the origin of the pulmonary artery, at a point on a level with the lower border of the left third costal cartilage, just behind the left edge of the sternum. It passes obliquely upward to the right, a little beyond the right edge of the sternum, in the right second intercostal space, to the upper border of the right second costo-sternal articulation. A needle passed through the second intercostal space close to the right edge of the sternum would, after passing through the lung, enter the pericardium and the most prominent part of the bulge of the aorta (Gray). This second intercostal space on the right side of the sternum is therefore also called the aortic interspace. The venous blood, emptying into the right auricle from the superior and inferior venæ cavæ, passes through the tricuspid orifice, into the right ventricle. The direction of the blood at first is toward the apex, but it suddenly curves upward to the left, and is driven by ventricular systole through the pulmonary orifice, and by the pulmonary artery it is conveyed into the lungs, for aëration. From the lungs the aërated (arterial) blood is conveyed by the pulmonary veins to the left auricle. Thence through the mitral orifice into the left ventricle. Here the blood current, as in the right ventricle, is directed at first toward the apex, but immediately curves upward to the right, to the aortic opening, through which it is driven by ventricular systole. Closure of the mitral and tricuspid valves occurs with ventricular systole, and prevents regurgitation from the ventricles into the auricles; and closure of the semilunar (sigmoid) valves guarding the pulmonary and aortic orifices occurs with ventricular diastole (arterial systole), to prevent regurgitation from the arteries into the ventricles.

Situation of the Valves.—The pulmonary valves are situated highest up in the thorax of any of the valves of the heart. A needle pushed through the centre of junction of the left third costal cartilage with the sternum, would penetrate about the centre of the pulmonary orifice. We do not, however, listen directly over this point for sounds connected with the pulmonary orifice in their loudest intensity, for the bone intervenes; but we listen in the second left, or pulmonary (pulmonic), interspace, where the sounds are conveyed.

The aortic valves are situated behind the pulmonary, a little lower down and to the right, just behind the left edge of the sternum on a level with the lower border of the third rib. We do not listen here through the bone for aortic sounds, but in the second right, or aortic, interspace where they are conveyed.

The mitral valves guarding the orifice between the left auricle and left ventricle, are situated deeply within, at a point corresponding with the upper border of the left fourth costal cartilage, near the left edge of the sternum. We do not listen here for sounds connected with the mitral orifice, for the right ventricle and

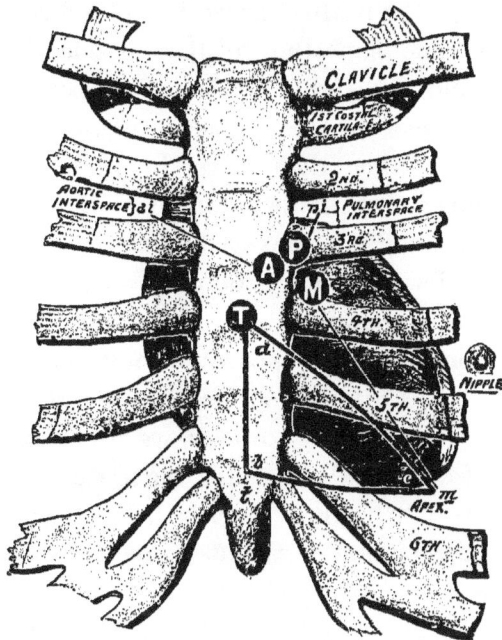

Fig. 20.—Diagram showing Location of the Valves of the Heart, and Points of Maximum Intensity of Sounds connected with them. The triangle a b c is the area of superficial dullness.

pulmonary tissue are in front of it; but we listen down at the apex, which is made of the left ventricle, and to this point mitral, as well as other left ventricular, sounds are conveyed.

Lastly, the tricuspid valves are situated behind the median line of the sternum, between the fourth costosternal articulations. But we do not listen at this point for tricuspid sounds, but at the point where the lower border of the right ventricle crosses the sternum, about the base of the ensiform cartilage. The tricuspid valves guard the orifice between the right auricle and right ventricle.

A circle of one inch in diameter includes parts of all the valves of the healthy heart, but of course they are not in the same plane, those of the left side of the heart being behind the right. It is very important to observe, also, that we do not, as a rule, listen directly over the orifices and their valves in order to best hear sounds connected with them, but over those points to which such sounds are conveyed with greatest intensity, as follows: for pulmonary sounds, over the pulmonary (second left) interspace; for aortic sounds, over the aortic (second right) interspace; for mitral sounds, over the apex; and for tricuspid sounds, over the ensiform cartilage. Posteriorly, however, we do listen over the location of the mitral valves for the mitral regurgitant murmur, as will be fully described, as will also be the areas of transmission of various sounds.

Sounds of the Heart.—There are two sounds of the heart—first and second. As the first sound is heard loudest at the apex, it is also called the apex or inferior sound of the heart; and because it occurs during systole it is also called the systolic sound. The second sound is best heard at the base, and hence is sometimes called the basic or superior sound of the heart, and because it occurs in diastole it is also called the diastolic sound.

The first sound (inferior, apex, systolic) of the heart is a composite sound, partly due to closure of the mi-

tral and tricuspid valves, and partly due to the apex-
beat, the rush of blood, and the stretching of the
chordæ tendineæ, to say nothing of other elements.
But whatever elements take part in its production, it
is necessary, and very important, to know and remember that the first sound in the normal heart is synchronous with (occurs at the same time with) the closure
of the mitral and tricuspid valves, the systole of the
ventricles, and the apex-beat. The latter slightly precedes, of course, the radial pulse. This first sound,
though heard almost at any part of the chest in
some cases, is best heard in all at the apex, as already
stated, and sounds like *ub*, in the words tub or rub.
But it does not sound like *rub* or *lub*, except when
there is a præsystolic murmur, as we shall see. It is
longer in duration and lower pitched than the second
sound.

The second sound (superior, basic, diastolic) of the
heart, is produced by the simultaneous closure of the
semilunar (sigmoid) valves of the aortic and pulmonary
orifices, and is synchronous with diastole of the ventricles. It is heard best at the base of the heart, and is a
shorter, sharper, and higher-pitched sound than the
first, and resembles the word *up*, in cup.

Rhythm of the heart is the repetition of all the successive phenomena which go to make up what is termed
a complete circuit or revolution, each one of which is
divided into a first sound, first rest, second sound, and
second rest.

Suppose a revolution to be ten-eighths of an inch
long: the first sound would be, according to Walshe,
four-eighths (half-inch), the first rest one-eighth, the

second sound two-eighths (quarter of an inch), and the second rest three-eighths of an inch long, thus:

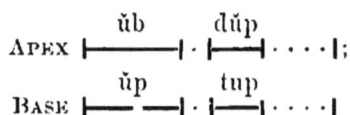

ŭb   dŭp

APEX ├────┤ · ├───┤ · · · ·│;

ŭp   tup

BASE ├── ──┤ · ├──┤ · · · ·│

FIG. 21.—Normal Rhythm of the Heart as heard at the Apex and Base.

The heart's rhythm may be imitated by striking a table, for instance, with the palmar surface of the hand, near the wrist, for the first sound, and with the point of the finger for the second, observing the proper intervals for the periods of silence. In the accompanying diagram the consonants *d* and *t* are placed before the second sound, as heard at the apex and base respectively, merely for the sake of euphony, and not because there are really any such elements of sound in the normal rhythm of the heart.

FIG. 22.—Sphygmographic Tracing—Normal Heart. (Walshe.)

We now proceed to consider the heart in its various abnormal conditions.

VALVULAR LESIONS OF THE HEART.

Valvular lesions commonly give rise to enlargement of the heart. They usually result from one or more previous attacks of endocarditis. The latter disease frequently occurs in the course of acute articular rheumatism, especially among the young, but it may also occur during an attack of diphtheria, scarlet fever, typhoid fever, measles, syphilis, lead poisoning, gout,

PLATE VII.

FIG.A.

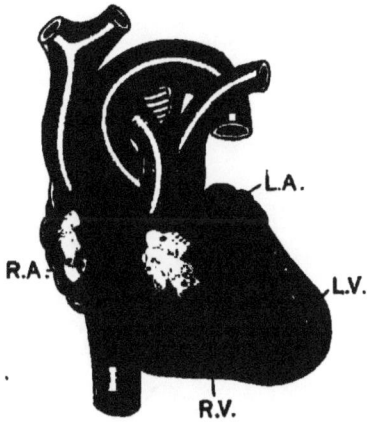

NORMAL HEART.

FIG.B.

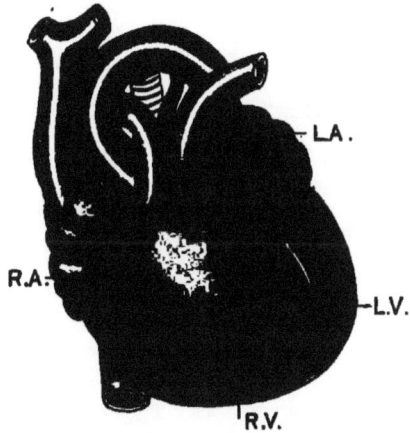

MITRAL REGURGITATION.— Showing
enlargement of LA.LV & RV.

FIG.C.

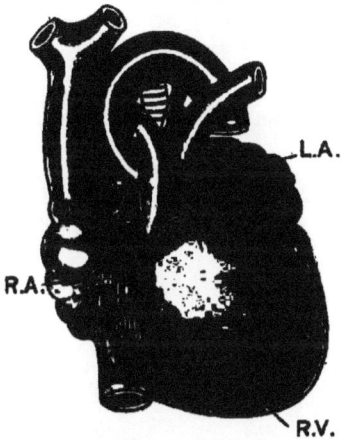

MITRAL OBSTRUCTION.— Showing
enlargement of LA & RV.

FIG.D.

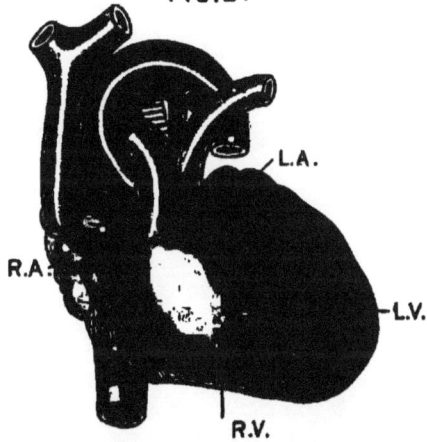

AORTIC OBSTRUCTION & REGURGITATION.
Showing enlargement of LV.

REFERENCE.

R.A. Right Auricle.
LA. Left —ıı—

R.V. Right Ventricle.
L.V. Left —ıı—

DRAWN BY H.MACDONALD M.D. N.Y.

erysipelas, Bright's disease of the kidneys, and other
diseases, or it may be due to pyæmia, or surgical injury, or, finally, it may occur independently. In the
latter case it is termed idiopathic endocarditis. Such
cases are, however, very rare. Valvular lesions, especially aortic insufficiency, may also be due simply
to violence, as in lifting. Endocarditis during fœtal
life attacks the right side of the heart, because, in that
state, the right side of the heart has more work to do
than the left. For the same reason endocarditis attacks
the left side of the heart after birth. It is extremely
improbable, that, as Richardson, of London, states, the
blood receives a poison in the lungs after birth, which
it takes directly to the left heart, causing endocarditis
of the left side of the heart; but by the time it gets
back to the right heart, the poison has lost its virulence.
If it be due to poison only, the right heart should be
poisoned instead of the left, for the blood is supposed
to be purified in the lungs, instead of being poisoned
there. Endocarditis does not necessarily leave traces
behind, but it usually does. It may, however, affect
the valves or orifices, or, they remaining intact, it may
leave some lesional traces on the wall within the ventricle at some point, or points.

From what has been said of fœtal endocarditis, we
conclude that children with valvular lesions of the tricuspid or pulmonary valves were born with them,
though they may not have been discovered for several
years afterward, when the changes in the heart produced would make the signs more observable. But
relative (or secondary) insufficiency of the tricuspid
valves occurs in persons after they are born. This is

13

due, not to inflammation, but to enlargement of the right ventricle (dilated hypertrophy) as in general emphysema, and mitral obstruction or regurgitation, where, owing to the dilated hypertrophy of the right ventricle, the tricuspid valves presently fail to close the orifice, on account of their being mechanically separated too widely.

Order of Frequency of Valvular Lesions.— All agree that mitral regurgitation is the most common. Regarding mitral obstruction authors disagree. Dr. Walshe places it last of the valvular lesions in the left heart, but I am disposed to think it is often present with regurgitation. The following is the order of frequency, according to Dr. Walshe:

(1) Mitral regurgitation, (2) aortic obstruction, (3) aortic regurgitation, (4) mitral obstruction, (5) tricuspid regurgitation (relative included), (6) pulmonary obstruction (most frequent inflammatory of the right heart), (7) pulmonary regurgitation (very rare), and (8) tricuspid obstruction (hardly known). I, myself, have never observed a case of pulmonary regurgitation, and tricuspid obstruction is, probably, only recognizable on *post-mortem* examination, if it should ever occur.

Valvular lesions, instead of existing singly, may be, and often are, combined.

Order of Frequency of Combinations.—According to Walshe they are as follows:

(1) Mitral regurgitation and aortic obstruction, both giving rise to systolic murmurs. (2) Aortic obstruction and regurgitation. (3) Mitral regurgitation and aortic regurgitation. (4) Mitral regurgitation, aortic obstruction and regurgitation. (5) Mitral obstruction and re-

gurgitation, and so on. From the foregoing, it will be observed that obstruction and regurgitation may, and do, exist at the same orifice. This is perfectly true, for while the orifice may be constricted, the valves may be prevented from closing by being fixed open by adhesion. As already mentioned, in a general way, valvular lesions usually produce enlargement of the heart. It may now be further stated that each valvular lesion is followed by enlargement, peculiar to itself. This is of the greatest importance in making a diagnosis, and should never be lost sight of by the examiner. In speaking of enlargement, also, not only is dilatation, or hypertrophy, meant, but both—dilated hypertrophy, or hypertrophous dilatation.

Now as to whether dilatation occurs first and hypertrophy afterward, authors again disagree. It seems reasonable that the two should proceed together, until the time arrives when hypertrophy ceases. Then uncompensated dilatation alone remains. That is the usual, inevitable tendency, if the patient lives long enough and does not die meantime of some complication or fatal intercurrent disease.

The particular enlargements (hypertrophous dilatations, or dilated hypertrophies) following the various lesions respectively, will be considered with each case of valvular disease.

Cardiac Murmurs.

Cardiac murmurs are adventitious sounds heard in connection with the heart in addition to, or in the place of, those sounds that exist in health. When due to organic disease they are termed organic murmurs.

But when due to anæmia or perverted cardiac action they are said to be inorganic, or simply functional. Either of these two classes of murmurs may originate without or within the heart, the former being termed pericardial (exocardial), the latter endocardial murmurs. They may also exist together or separately.

ENDOCARDIAL MURMURS.

Valvular lesions usually give rise to enlargement of the heart, and are generally accompanied by permanent murmurs. The latter assist greatly in making a correct diagnosis in each case, according to their location, areas of conduction and transmission, and also by their rhythm or time of occurrence with regard to the sounds of the heart. The properties, or elements, of murmur-sounds (quality, pitch, intensity, and duration) are also of some importance, particularly the quality, but these are secondary to other considerations, as will be seen. The loudness or feebleness of a murmur does not indicate the amount or gravity of the lesion giving rise to it. This is better told by the change in the form and size of the heart produced, and other considerations to be noted. The fact that a murmur is heard about the heart is no sign of itself that the heart is diseased at all, since there are many murmurs that are independent of actual cardiac disease, however closely they may imitate true organic cardiac murmurs. It is the business of the examiner to distinguish between them, as can usually be done by careful and intelligent observation.

There are other considerations, therefore, far more important than the mere fact of the presence of a murmur in connection with the examination of the heart.

Individual murmurs usually have certain points of maximum intensity, as well as areas of convection, conduction, or transmission, provided they are of the average intensity. But it sometimes happens that any murmur may be so loud as to be heard all over the body, whereas that which is usually the loudest murmur may be, or become, so feeble as to be heard with difficulty, if at all, at its point of maximum intensity. We will now consider the murmurs and other physical signs characteristic of valvular lesions, in the regular order of their occurrence.

MURMURS HEARD LOUDEST AT THE APEX.—MITRAL MURMURS.

We have already shown that there are four points on the front of the chest wall where the various heart-sounds and murmurs may be heard respectively in their maximum intensity. Mitral murmurs, for instance, are heard loudest at the apex, tricuspid murmurs over the ensiform cartilage, aortic murmurs usually over the aortic interspace, and pulmonary murmurs over the pulmonary interspace.

Omitting pericardial and pleuro-cardial friction sounds, to be presently considered, there are five murmurs heard only, or loudest, at the apex, and are consequently referable to the mitral orifice and left ventricle. Four of these murmurs are systolic in time, and one præsystolic. The four systolic murmurs are the mitral regurgitant, intra-ventricular, dynamic, and cardio-respiratory. Of these, the mitral regurgitant and intra-ventricular murmurs are organic, the dynamic and cardio-respiratory being inorganic, or functional.

The præsystolic murmur is also organic, being due to mitral obstruction. Of these five mitral apex murmurs, therefore, three are organic and two are inorganic, or functional. Compare the following table:

Mitral or Apex Murmurs	Systolic	Mitral Regurgitant Intra-Ventricular.
		Dynamic { Neurotic origin. / Due to Anæmia.
		Cardio-respiratory.
	Præsystolic . .	Mitral Obstructive.

MITRAL REGURGITATION.

Mitral regurgitation (reflux, insufficiency) is the most frequent of all valvular lesions of the heart. It is usually a primary result of endocarditis, although it may be relative or secondary to aortic regurgitation, or even marked aortic obstruction, which might lead to such dilated hypertrophy of the left ventricle that the mitral valves would become insufficient. Such cases are, however, comparatively rare. and, as already said, it is nearly always the direct result of endocarditis affecting the mitral valves themselves.

Mitral regurgitation leads to the following change in the form of the normal heart: (1) enlargement (dilated hypertrophy) of the left auricle; (2) enlargement of the left ventricle. and (3) enlargement of the right ventricle.

It is easy to understand why the left auricle and right ventricle become enlarged. The blood regurgitating back into the left auricle during ventricular systole, instead of going on through the aortic orifice, gives the left auricle increased work to do. and it necessarily becomes enlarged (dilated hypertrophy). In the same way, the blood forced back on the lungs gives the right ventricle more work to do in driving blood

through the lungs. But just why the left ventricle becomes enlarged also, the mechanism is not so clear.
As Walshe says, the enlargement of the left ventricle
in these cases "is plainly supplemental; though it is
difficult to see how a condition that makes regurgitation more forcible, can tend to balance the evils arising
from it." In regard to this point, my attention was
called by my clinical assistant, Dr. William C. Rives,
to Vierordt, of Leipsic, who, says truly, that "the left
ventricle first becomes dilated from having blood under
abnormally high pressure, and increased in quantity,
driven into it during its diastole by the enlarged left
auricle. The ventricle then becomes hypertrophied in
order to dispose of this extra quantity of blood, partly
forward into the aorta, and partly backward into the
left auricle." That seems to be the correct view of the
matter, but whatever theories there may be about it,
one thing is certain, and that is that in mitral regurgitation the left ventricle becomes enlarged in some way
from overwork.

Another point is not to be overlooked. The second
sound of the heart is louder, usually, than normal over
both the pulmonary and aortic interspaces (second interspaces on the left and right side of the sternum respectively). But as heard over the pulmonary interspace it is often somewhat louder (accentuated) than
that heard over the aortic interspace, the latter being
the weaker of the two, owing to regurgitation at the
mitral orifice. The reason for this accentuation of the
second sound of the heart over the pulmonary interspace is obvious. Owing to enlargement (dilated hy-

pertrophy) of the right ventricle, the blood is thrown
with increased force into the pulmonary artery, and
owing to increased tension in that vessel, the valves
close more forcibly, until relative tricuspid insufficiency
occurs. This phenomenon is still more marked in mitral obstruction, as we shall see.

The radial pulse will be found also in mitral regurgitation to be irregular in size, sometimes large, sometimes small, and generally compressible. It may also
intermit, but irregular rhythm is not peculiar to any
valvular lesion, as we shall see when speaking of irregular rhythm in general (p. 264).

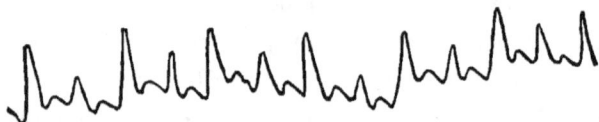

Fig. 23.—Diagram of Sphygmographic Tracing of Pulse in Mitral Regurgitation.
(Walshe.)

In the early existence of mitral regurgitation, before
enlargement has had time to occur, as in other lesions,
or toward the end, when there are dilation and feebleness of action, with perhaps pulmonary infarctions and
other complications that are likely to arise, the physical signs are not usually so clear. But in general they
are as follows:

Inspection.—The apex-beat is usually visible, owing
to its force from enlargement, and is observed to be
displaced downward and outside of the mammillary
line, indicating enlargement of the left ventricle. This
is of the utmost importance in the diagnosis of this
disease. The heart's impulse is usually seen to be more

forcible than normal. Sometimes, in persons with thin chest walls, left auricular impulse is observed in the pulmonary (left second) interspace, and will be one of two kinds: (1) systolic, if communicated to the auricle from the ventricle; or (2) præsystolic (auriculo-systolic), if due to hypertrophy of the auricle. As the case progresses, jugular pulsation on the right side of the neck may be observed, due to relative tricus-

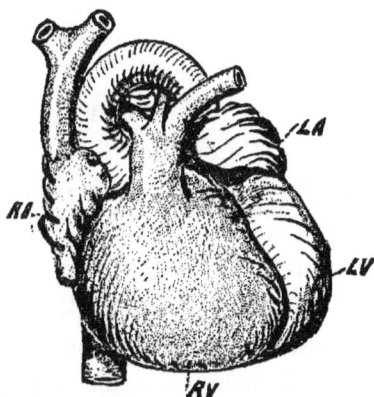

Fig. 24.—Diagram of the Heart in Mitral Regurgitation. Left Ventricle, Left Auricle, and Right Ventricle are seen to be Enlarged—compare normal heart.

pid insufficiency from enlargement of the right ventricle.

Palpation.—The apex-beat will be felt downward and outward away from its normal position, and the heart's impulse will be felt to be generally increased in force. Pulsation near the ensiform cartilage is sometimes seen and felt, and is due to enlargement and forcible action of the right ventricle. The latter sign is not so well marked here as in general emphysema, when the right ventricle alone is enlarged, and the whole heart is also pushed down by the increased volume of the lungs.

Purring Thrill.—In some cases of mitral regurgitation, a systolic purring thrill is distinctly felt, especially if the palm of the hand is placed very lightly over the lower præcordium. It is usually felt in the fourth interspace, on or near the mammillary line. Purring thrill here, as in other cases of valvular disease, depends for its production upon three abnormal factors: (1) increased capacity of the heart, (2) increased propelling force, and (3) an abnormal orifice. It will not be produced by increased force and an abnormal orifice alone. Besides hypertrophy, there must also be enlargement of the cavity of the heart, with a certain amount of thinness of its walls, as seen after dilatation has occurred to some extent. With increased capacity and force, and a stream of blood driven through a button-hole orifice, for instance, thrill is almost certain to occur. But it may also be due in some instances, I think, to a ribbon-like vegetation, with one end free and vibrating in the course of the blood current. However this may be, thrill is not a constant phenomenon. It will disappear from changes in the orifice, or after hypertrophy ceases to compensate, and the walls of the heart become weak. The disappearance of thrill, therefore, after it has once been present, would seem to be not always a favorable sign.

Percussion.—The areas of both deep and superficial dullness are enlarged, the latter especially, downward and to the left. Dullness over the left auricle, especially among those with thin chest walls, is more marked and extends beyond the normal limits already mentioned.

Auscultation.—A blowing systolic murmur is heard, but loudest at the apex, for reasons already given (p. 189), and the second sound of the heart is usually more or less accentuated over the pulmonary (left second) interspace, since the enlarged right ventricle would cause greater tension in the pulmonary artery, with consequently more forcible closure of the pulmonary valves than would be the case in the aorta. After tricuspid insufficiency occurs, this accentuation is not so marked. This murmur has various other names, such as mitral systolic, mitral indirect, and mitral insufficient.

The mitral regurgitant (systolic, indirect, insufficient) murmur is, as already stated, usually blowing in quality, and occurs with, or takes the place of, the first (systolic) sound of the heart. It also occurs, therefore, with the apex-beat. By paying attention to the last named point, the time of the murmur can usually be fixed, even if the heart's rhythm be irregular. The following diagram represents the mitral regurgitant murmur:

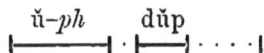

$$\text{ŭ-}ph \qquad \text{dŭp}$$
$$\vdash\!\!-\!\!-\!\!-\!\!\dashv \cdot \vdash\!\!-\!\!\dashv \cdots \dashv$$

Musical Murmurs.—Sometimes these and other murmurs are musical, instead of possessing a blowing, blubbering, rough, or other quality. This quality is of no particular import, but simply indicates that "prominent spiculæ or fibrinous particles, of vibratile character, project into the current, or else that rigid vibratile edges bound a narrow, chink-like opening" (Walshe).

Area of Transmission.—If the murmur be a very
feeble one, as may occur in cases of long standing, with
marked dilatation of the heart and feebleness of its
action, especially if the chest walls be thick and fleshy,
it may be heard only at the apex, and then with difficulty. But if it be heard anywhere it will usually be
at the apex, which is composed of the left ventricle,
the right ventricle being in front of the body of the
left. But if the mitral regurgitant murmur be a very
loud one, especially if it be musical, as sometimes happens with any murmur, as already described, then it
may not only be heard at the apex, but all over the
chest. The average mitral regurgitant murmur may,
however, besides being heard loudest at the apex, be
also heard over three other localities: (1) posteriorly,
between the inferior angle of the scapula and body of
the eighth dorsal vertebra, or thereabouts; (2) along the
left lateral base of the chest; and (3) over the left auricle and to the left of it. In the first case, we are listening directly over the site of the mitral valves, and this
is perhaps the only case in which we listen directly
over the site of the valves of the heart for sounds in
connection with them. Though usually heard behind
at that point, it is not necessarily heard there. The
murmur may be too weak, as already stated; or the
patient may have an emphysematous lung interposed;
or, as in the case of a lady from Canada, whom I examined, the murmur was distinctly heard behind, until
accidentally falling ill with pleurisy with effusion, the
murmur disappeared posteriorly, and though she made
a fair recovery from the pleurisy, the murmur never
again returned to that point. Enlarged bronchial

glands and other acoustic impediments evidently might prevent such a murmur from being heard behind.

(2) Besides hearing the mitral regurgitant murmur at the heart's apex, and posteriorly over the mitral valves, it is sometimes, not always, transmitted from the apex along the ribs to the left; but it is not so transmitted to the point behind that has been mentioned. Why is it that this murmur is sometimes heard transmitted along the ribs to the left, and sometimes not? There are two classes of causes. First and chiefly, if the right ventricle becomes very much enlarged, it acts as a wedge between the left ventricle and the chest wall, and pushes the left ventricle back so far that it is impossible for the ribs to take up the sound. In these cases it will be very distinct posteriorly, not so much as is usual at the apex, and not heard at all to the left. Such conditions would indicate a very large right ventricle. The second class of causes would be pleurisy with effusion, thickened pleura, emphysema, or some other acoustic impediment.

(3) Lastly, if the left auricle be greatly enlarged and the murmur be communicated to it, the murmur may be heard in the third and even second, interspace on the left of the sternum, over the site of the left auricle, and thence transmitted to the left axilla.

Diagnosis of Mitral Regurgitation.—If the existence of mitral regurgitant murmur can be established, the diagnosis is complete. But the extent of the lesion here, as elsewhere, cannot be accurately estimated by the properties of the murmur, but by the amount of cardiac enlargement produced.

Pericardial (exo-cardial) friction sounds are super-

ficial, rubbing, churning, grazing or creaking in quality,
are not transmitted beyond the limits of the heart,
change in intensity with position of the patient in
leaning forward or backward, or by pressure with the
stethoscope, and have no fixed relation in time to the
heart-sounds.

Pleuro-cardial, or pleuritic friction sounds, near by,
may be kept up by the heart's impulse, and do not
necessarily cease upon holding the breath, but are easily
distinguished by their quality and the circumscribed
area to which they are limited. There remain three
other systolic apex murmurs, which may very closely
imitate the mitral regurgitant, and it is necessary to
know how to distinguish them. They are (1) intra-
ventricular murmurs, termed also by Flint and others
the mitral systolic non-regurgitant murmur, (2) dynamic, and (3) cardio-respiratory murmurs. Intra-ventricular murmurs are organic, whereas the dynamic
and cardio-respiratory murmurs are inorganic, or functional.

(1) *Intra-ventricular murmurs*, when they exist,
are usually so feeble that they may be heard at the
apex only, or base. In either case they are always
systolic, and due to lesions somewhere within the ventricle, instead of affecting the orifices or valves. Some
of the intra-ventricular murmurs are heard only at the
apex, others at the base, others again more or less over
the whole heart. They are due to roughening of the
ventricular endocardium, misattachment of a chord,
fibrinous shreds across the blood current, thickening
and roughening of chords, twisting of columnae carneae,
inflammatory vegetations on the wall of the ventricle,

and cardiac ventricular aneurism. In the latter case the ventricle is enlarged, and it may be impossible to make a diagnosis; but in all other cases of intra-ventricular murmurs the diagnosis is easy, since the cause of their production is not a cause for enlargement or perceptible interference with the circulation. The murmurs of this variety are usually weak, and for that reason only are restricted to a spot over the apex, and are not transmitted anywhere, or heard posteriorly. The prognosis in the two cases would be entirely different. For whereas mitral regurgitation is usually fatal in course of time, owing to pulmonary congestion, secondary tricuspid insufficiency, hemorrhagic infarctions of the lungs, and cardiac dropsy, intra-ventricular lesions may be of little or no importance, though sufficient to produce a murmur.

(2) *Dynamic Murmurs.*—These murmurs are due to some perverted action of the heart, and are observed sometimes in choreic subjects. The heart is a voluntary muscle, and hence subject to choreic, irregular movements, like other voluntary muscles. Consequently, during systole, some of the columnæ carneæ twitch and pull the mitral valves open, causing a faint murmur, which is only audible at the apex. Not only so, but it is not necessarily associated with left ventricular, or other cardiac enlargement, and, moreover, the dynamic murmur from this cause is very inconstant, dependent, as it is, upon choreic movements which are absolutely uncertain. It is sometimes found, also, in those who are nervous from other causes, as among hysterical women, tobacco smokers, and rarely among those nervous from abuse of alcohol. It is also

heard sometimes in epileptic subjects. But in all such
cases the murmur is weak, and hence limited to the
apex usually, is not necessarily attended with enlargement of the heart, and, above all, is inconstant.

Dynamic murmurs, instead of being of neurotic origin, may also be due to anæmia. No truly hæmic
murmur is ever heard at the apex, unless produced by
heart clot, which is very rare. But hæmic murmurs
due to the watery condition of the blood are heard only
at the base and over the pulmonary interspace, unless
caught up by the aorta, and are caused by watery
venous blood, as will be considered fully when speaking of basic murmurs. That anæmia does give rise
indirectly, or dynamically, to a systolic apex murmur,
besides causing all cardiac murmurs to be louder than
they would be otherwise, by increasing vibratility of
tissues, cannot be denied. But such a systolic apex
murmur should not, strictly speaking, be termed an
anæmic murmur, but a dynamic murmur due to anæmia. The mode of its production is as follows: The
heart becomes flabby, and the papillary muscles become
weak from fatty degeneration, according to Guttmann
and others, thus allowing the mitral valves to recoil
too far, so as to cause a slight backward leakage; or
else temporary dilatation from anæmia may allow a
vortiginous (eddying) movement of the blood within
the ventricle, or misdirection of its current, sufficient
to give rise to a systolic murmur.

The masturbator's systolic apex murmur, described
by Walshe, and attributed by him to nervous influence, may also be due to the anæmic condition of the
patient.

By curing such patients of their anæmia by means of iron and nutritious diet, these murmurs will disappear, it is true, but even then they are not to be regarded as true hæmic murmurs, though venous hum be also present, but as dynamic murmurs due to the anæmic condition. The venous hum in the neck, with which they are usually associated, will be fully considered in connection with basic heart murmurs.

(3) *Cardio-respiratory murmurs* are always systolic, and heard at the end of a full inspiration, as the heart's impulse may force air out of some vesicles or cavities near by, with a sound that sometimes imitates the cardiac systolic murmur; or the respiratory murmur itself may sound like a heart murmur during systole. Simply have the patient hold the breath after expiration, and the imitation heart murmur at once ceases. It is always a good rule, during auscultation of the heart, to have the patient hold the breath at least once for each point examined.

MITRAL OBSTRUCTION.

Mitral obstruction (constriction, stenosis) is said by some authors to be idiopathic, or congenital, and not traceable to previous rheumatic endocarditis, or other disease. In these cases it chiefly affects women and children. But it not infrequently does result from rheumatic endocarditis in the young of both sexes. Aortic, rather than mitral lesions, are apt to occur after middle life.

Mitral obstruction leads to enlargement (1) of the left auricle, and (2) of the right ventricle, and by en-

largement is meant here, as elsewhere, dilated hypertrophy.

The difference in enlargement, therefore, in mitral
obstruction and regurgitation, is the left ventricle. In
obstruction it is not only not enlarged, but, on the contrary, somewhat smaller than normal, from diminished
volume of blood entering it; but in mitral regurgitation it is enlarged, as already mentioned. In mitral
obstruction the left auricle becomes enlarged, from the

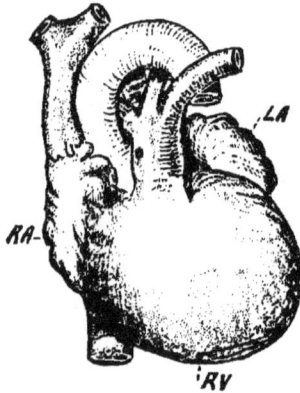

Fig. 25.— Enlargement of the Left Auricle and Right Ventricle in Mitral Obstruction.

effort to drive blood through an obstructed orifice, and
the blood, being prevented from leaving the lungs
freely, the right ventricle becomes enlarged from the
extra task of driving the blood through them. The
fact that the left ventricle is not enlarged in mitral obstruction is of the greatest importance in making a
correct diagnosis. The second sound of the heart is
usually accentuated in the pulmonary interspace, for
reasons already given (p. 199), and weakened in the
aortic interspace, for the reason that the left ventricle

is somewhat atrophied, with consequent diminished tension in the aorta. The difference is more marked here than in mitral regurgitation.

In addition to this accentuation, the second sound is also reduplicated at the base in about one third of all cases, as the pulmonary valves close not only more forcibly than the aortic, due to increased tension in the pulmonary artery, from enlargement of the right ventricle, but also earlier than the aortic valves, for the same reason.

The radial pulse is not noticeably affected in mitral obstruction, nor is the heart's rhythm necessarily disturbed, but is generally regular. The physical signs of mitral obstruction are:

Inspection.—The apex-beat, if seen at all, will usually be within the mammillary line, as the left ventricle is not enlarged. It may be pushed out a little, however, by enlargement of the right ventricle. The apex-beat will be observed to be not more forcible, perhaps, than in health. But there may be observed in those having thin chest walls a left auricular systolic impulse over the left third interspace, owing to enlargement of the left auricle. This impulse immediately precedes the apex-beat, and with regard to the latter is præsystolic. Pulsation of the enlarged right ventricle will not be observed usually, unless the heart be lowered from some cause, when it will be observed near the end of the ensiform cartilage, as in general vesicular emphysema.

Mitral obstruction occurring in children appears to be not infrequently associated with the so-called pigeon-breast. The flattening is especially well-marked in the lower præcordial region on the left of the sternum. Whether this is mere coincidence in a certain number of cases, or due to atrophy of the left ventricle, or to lack of nutrition in general, from imperfect cardiac function, is not exactly known.

As the case progresses and relative insufficiency of the tricuspid valves occurs, due to excessive dilated hypertrophy of the right ventricle, jugular pulsation on the right side of the neck, and afterward also on the left side, will be observed. Then follows cardiac dropsy, commencing in the feet.

Palpation.—The apex-beat may be felt to be not markedly displaced or increased, and there may be some pulsation near the end of the ensiform cartilage, due to enlargement of the right ventricle, if the heart be sufficiently lowered. But what is most distinctive is the præsystolic thrill often felt about the left fourth interspace. By placing the palm of the hand lightly over the part, the thrill may be felt immediately before the apex-beat, and it is characteristic of mitral obstruction, though not always present. It is due to the forcible contraction of the enlarged left auricle in its endeavor to force the blood through an obstructed orifice. The general mode of production is the same here as in regurgitation (which see, p. 202) and elsewhere, and, like that, is not permanent. Auricular impulse, præsystolic in time with regard to the apex-beat, is sometimes felt.

Percussion.—The area of dullness over the left auricle and right ventricle is enlarged, but what is known as the superficial area of cardiac dullness is not so much enlarged as in mitral regurgitation, since in mi-

tral obstruction the left ventricle is not enlarged, and
a small part of the superficial area of dullness is normally formed of the left ventricle.

Auscultation.—A blubbering præsystolic murmur,
like vibrating the letter *r* with the tongue, or vibrating the flaccid lips by blowing forcibly (expiration)
through them while they are closed (Flint), is usually
heard loudest at the apex, and limited to that region,
though it may be conveyed up to the fourth interspace
by the blood current, and in some rare cases is so loud
as to be heard behind—in fact, all over the chest. The
second sound of the heart is usually accentuated over
the pulmonary interspace. The murmur has various
names, such as præsystolic, direct, constrictive, stenotic,
and so on. As the murmur in regurgitation is termed
regurgitant, so in obstruction it may be called obstruct-
ant or obstructive.

The mitral obstructive (direct, stenotic, constrictive,
præsystolic) murmur is usually blubbering in quality,
as already stated, and unless it is blubbering it is usually absent altogether, for this is the only organic heart
murmur that may appear and disappear. All other
organic heart murmurs are permanent. The reason
why it is blubbering, as Flint explains, is because of
the vibration of the free edges of the valves, the orifice
between them being narrow and bottonhole-like. It is
like throwing the flaccid lips into vibration, by forcibly
expelling the breath while the mouth is gently closed,
which, as Flint truly states, "represents not only the
characteristic quality of the murmur, but the mode of
its production." When the edges of the mitral valves,
instead of being flaccid, become fixed from inflamma-

tion, the murmur ceases and regurgitation follows. It also ceases if the circulation is feeble, but in that case it may return with thrill on exercising. The following diagram represents the mitral obstructive murmur:

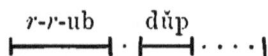

r-r-ub   dŭp

├────────┤ · ├────┤ · · · · |

This blubbering murmur at the apex, according to Flint, is sometimes due to aortic regurgitation, causing a secondary, or relative, mitral obstruction. In this case the edges of the healthy mitral valves are thrown into vibration slightly by the blood from the left auricle, while they are being closed by the backward pressure of blood due to aortic regurgitation. This is different from the aortic diastolic regurgitant murmur that is sometimes transmitted faintly to the apex.

The mitral obstructive murmur is never due to anæmia, or any other functional or inorganic cause. It is always an organic murmur, due usually to mitral obstruction primarily, and rarely to aortic regurgitation, which causes a secondary, or relative mitral obstruction. The idea that there is no such thing, in fact, as a mitral obstructive (præsystolic, direct, stenotic, constrictive) murmur, but that it is really regurgitant, originated, it appears, with Barclay. Dickinson and others have followed Barclay in this notion, but Gairdner, Balfour, Bristowe, Flint, Loomis, and many others have demonstrated quite clearly that Barclay and his followers were mistaken. For according to Bristowe, for instance, if the mitral obstructive (præsystolic, direct, stenotic, constrictive) murmur is really regurgitant, due to prolonged contraction of the ventricle, then in

aortic obstruction there ought also to be a præsystolic
aortic obstructive murmur, a clear case of *reductio ad
absurdum.* Moreover, *post-mortem* examination sets
the whole matter at rest in favor of the mitral præsystolic (obstructive) murmur.

This murmur was pointed out, in 1841, by Gendrin,
of Paris, who called it præsystolic. It was first claimed
to be due to mitral obstruction by Fauvel, of Paris, in
1843. In 1861, Gairdner, of Glasgow, named it left
auriculo systolic murmur. It is also sometimes called
post-diastolic. Guttmann calls it a diastolic murmur.
Flint, however, makes the mitral diastolic murmur to
be different from the mitral præsystolic murmur proper,
and says that it is caused by the rush of blood through
and over abnormal structures before the auricle contracts. It is simply a part of the same murmur, and
may be said to be a distinction without a difference.
It is simply the first part of a prolonged mitral obstructive murmur, which occupies the whole of diastole
instead of being merely præsystolic.

Area of Transmission.—The mitral obstructive
murmur is usually limited to the region of the apex,
or conveyed up the ventricle to the fourth interspace,
unless it is very loud, when, in some cases, it may also
be heard posteriorly. Why is not this murmur usually heard posteriorly at a point between the inferior
angle of the scapula and body of the seventh or eighth
dorsal vertebra, as in the case of mitral regurgitation?
Simply because the murmur, being made in the direction of the blood current, and not against it, is not held
at the mitral orifice long enough to be heard, but is at
once carried down to the apex, where it is heard. On

the other hand, the mitral regurgitant murmur is conveyed backward toward the ear, when auscultating
posteriorly, and it is held there long enough to be
heard. Why is not the mitral obstructive (præsystolic,
direct, stenotic, constrictive) murmur transmitted along
the chest walls to the left, as in the case of mitral regurgitant murmur? Simply because in mitral obstruction, the left ventricle is not enlarged and the right
ventricle, which does become enlarged, and is in front
of the left, wedges off the latter from the chest walls,
so that it does not come near enough for them to pick
up the sound of the murmur. The murmur also is
præsystolic, and occurs before the apex could reach the
chest walls, even if the enlarged right ventricle were
not between the apex and the chest walls.

Diagnosis.—Mitral obstructive murmur would not
be mistaken for mitral regurgitant, but that sometimes
the cardiac rhythm is so irregular, and the sounds may
so nearly resemble each other, that it is difficult to tell
the first sound from the second. In these cases, as well
as where the rhythm is perfectly regular, the changes
effected in the form and size of the heart are not to be
overlooked. For besides the fact that mitral obstructive murmur is præsystolic in time, limited to the
apex, and blubbering in quality, while mitral regurgitant is systolic in time, usually heard behind, as well
as at the apex, sometimes transmitted along the left
lateral base of the chest, and blowing in quality, it
must not be forgotten that in mitral obstruction the
left ventricle is not enlarged, while in mitral regurgitation it is enlarged, with corresponding displacement
of the apex-beat downward and outward, usually out-

side of the mammillary line. Aortic regurgitant
sounds, though sometimes conveyed to the apex, are
attended with enlarged left ventricle, besides being
purely diastolic in time, and are not heard loudest at
the apex.

A rather faint mitral præsystolic murmur may sometimes be heard, due to relative or secondary obstruction to the onward flow of blood in some rare cases of
aortic regurgitation. The mitral murmur in such cases
is not usually attended with appreciable enlargement
of the right ventricle, and accentuation of the second
sound in the pulmonary interspace, but there is great
enlargement of the left ventricle which does not occur
in mitral obstruction.

MITRAL REGURGITATION AND OBSTRUCTION.

This combination of lesions not infrequently exists,
for it is perfectly clear that while the mitral orifice is constricted the valves may be prevented from
closure by adhesion. In such a case there would be
both regurgitation and obstruction at the same orifice.
Usually, there is but one murmur in these cases, the
mitral regurgitant, since, as already remarked, the mitral obstructive murmur usually ceases when the free
edges of the valves become fixed and cannot vibrate.
Sometimes, however, both murmurs are heard, the præsystolic and systolic—the former at the apex and transmitted upward with the blood current toward the
fourth interspace, the latter also at the apex, but transmitted around toward the left, and sometimes heard
posteriorly. The left ventricle, the left auricle, and

right ventricle are enlarged, and the second sound is more or less accentuated in the pulmonary interspace.

TRICUSPID REGURGITATION.

Tricuspid murmurs are usually so feeble that they are audible, as a rule, only over the ensiform cartilage. And inasmuch as the tricuspid obstruction (praesystolic, direct, stenotic, constrictive) murmur is so rare that it may practically be thrown out altogether, there remains only the tricuspid regurgitant murmur to be sought for in connection with suspected tricuspid lesions.

Tricuspid regurgitation (reflux, insufficiency) is commonly secondary or relative to enlargement (dilated hypertrophy) of the right ventricle, due to mitral regurgitation or obstruction, or to general vesicular emphysema. In all of these cases, as we have seen, the right ventricle in time becomes enlarged from the extra work put upon it in order to drive forward an impeded pulmonary circulation. After a while, in some of these cases, not all, the tricuspid valves become so widely separated that tricuspid regurgitation with jugular pulsation and cardiac dropsy result.

Primary or actual tricuspid regurgitation sometimes is met with, and then it is the result of fœtal endocarditis, as has already been referred to. It leads to enlargement of the right ventricle. The tricuspid regurgitant murmur is systolic in time and blowing in quality, but is usually so feeble that, as has already been stated, it is confined to a small area over the ensiform cartilage. Sometimes the aortic regurgitant murmur is so transmitted down the sternum that it is heard only over the ensiform cartilage, but tricuspid regurgitant murmur is systolic, while aortic regurgitant is diastolic in time. The latter is also accompanied by enlargement of the left ventricle and other signs to be described. The mitral regurgitant murmur is systolic in time, but it is heard loudest at the apex and transmitted to the left, besides being usually heard posteriorly. It sometimes exists with tricuspid regurgitation, so that the two murmurs run into each other. In this case the other signs of tricuspid regurgitation are to be considered, such as jugular pulsation, cyanosis, cardiac dropsy and pulmonary œdema. The character of the radial pulse is not affected by valvular disturbances of the right side of the heart. The rhythm is, however, sometimes irregular, especially in general vesicular emphysema (p. 266). There are no intra-ventricular, dynamic, or cardio-respiratory murmurs to be considered usually in differentiating the tricuspid regurgitant murmur. Should any such case arise, the same rules hold good in regard to them, and pericardial, or pleuro-cardial, friction sounds, as in mitral regurgitation, to which the reader is referred.

Jugular Pulsation.—This phenomenon was first observed by Lancisi, of Rome, in 1728. It is usually systolic, but is sometimes also præsystolic.

Systolic jugular pulsation is accounted for as follows: In tricuspid insufficiency from any cause, venous blood is forced back during systole through the tricuspid orifice into the right auricle, and against the column of blood in the venæ cavæ. The superior vena cava, right vena innominata, and right internal jugular vein, form a sort of consolidated, straight, trunk line, so to speak, and there is no valve in the way until we come to the lower end of the internal jugular vein, just at the root of the neck. Consequently, this systolic pulsation of venous blood always first occurs on the right side at the root of the neck. After that valve gives way, the pulsation extends up to the second valve, and finally

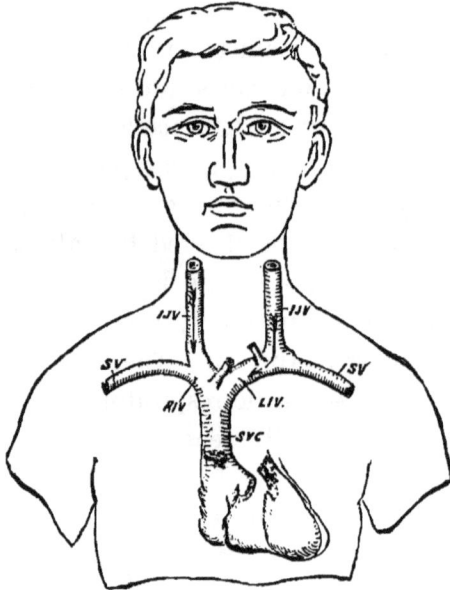

Fig. 26.—Jugular Pulsation. Position and Relation of the Veins of the Neck.

may reach the angle of the jaw. Meantime, it begins to appear also on the left side, but not so forcibly, as the backward shock has to go round a curve. In time, all the veins about the neck may participate in this curious phenomenon. Slight systolic jugular pulsation may, according to Bamberger, sometimes be observed on the right side, at the root of the neck, even when the tricuspid valves are not actually insufficient. In right ventricular enlargement a slight backward shock may even be imparted through the tricuspid valves themselves.

Præsystolic jugular pulsation is sometimes seen in cases of right auricular overflow, during contraction of the right auricle, and just before systole of the right ventricle, but it is rare. It sometimes occurs in tricuspid insufficiency, but tricuspid obstruction would also favor its production.

TRICUSPID OBSTRUCTION.

This lesion is very rare, and even when it is present there is usually no murmur heard with it. This is probably owing to the feebleness of the venous blood current as compared with the arterial, and the weakness of the right auricle as compared with the left. Usually, it is only on *post-mortem* examination that tricuspid obstruction is found to have existed.

Should there be any murmur present it would be præsystolic, as in the case of mitral obstruction, but the murmur would be limited to the region of the ensiform cartilage. The right auricle would be likely to become enlarged, but no particular change in the form and size of the heart is known. The radial pulse would be unaffected.

MURMURS HEARD LOUDEST AT THE BASE. AORTIC MURMURS.

We have seen that we listen over the apex and ensiform cartilage, respectively, in order to hear mitral and tricuspid murmurs in their maximum intensity. We

now come to a third point on the front of the chest
walls, which is the aortic interspace, or the right second
interspace, being that between the second and third
costal cartilages, on the right of the sternum. Here we
listen for murmurs referable to the aortic orifice in
their maximum intensity (pp. 187, 188).

Omitting pericardial and pleuro-cardial friction
sounds, already described under mitral regurgitation
(p. 198), there are seven basic murmurs heard only, or
loudest, over the aortic interspace, and are consequently
referable to the aortic orifice, the aorta itself, or the
left ventricle. Five of these murmurs are systolic in
time, and two diastolic. The five systolic murmurs
are the aortic obstructive, intra-arterial, intra-ventricular, dynamic, and cardio-respiratory. Of these, the
aortic obstructive, intra-arterial, and intra-ventricular
murmurs are organic, the dynamic and cardio-respiratory being inorganic, or functional. The diastolic murmurs in this case, as in others, are always organic. Of
these seven aortic basic murmurs, therefore, five are
organic and two are inorganic, or functional. Compare
the following table:

Aortic Basic Murmurs	Sytolic	Aortic Obstructive. Intra-Arterial. Intra-Ventricular.	
		Dynamic	Neurotic Origin. Due to Anæmia.
		Cardio-respiratory.	
	Diastolic	Aortic Regurgitant. Intra-Arterial (aortic non-regurgitant).	

By comparing this table with that of mitral apex
murmurs (p. 198), we have here, in addition, the intra-
arterial murmurs. We also have the aortic systolic

obstructive murmur in place of the mitral systolic regurgitant murmur, and the aortic diastolic regurgitant
instead of the mitral præsystolic (or diastolic, as it is
called by Guttmann) obstructive murmur. In other
words, the mitral regurgitant and aortic obstructive
murmurs are systolic, while the mitral obstructive and
aortic regurgitant are chiefly diastolic in time.

AORTIC OBSTRUCTION.

Aortic obstruction (constriction, stenosis) is the cause
of enlargement of only the left ventricle, as a rule. It
is the least harmful of all valvular lesions. The reason
for enlargement of the left ventricle in this disease is
obvious, as it has more work to do to drive blood
through the obstructed aortic orifice, and the degree of
enlargement will be in proportion to the amount of obstruction. The second sound over the pulmonary interspace remains normal, as the right ventricle is not affected, but the second sound over the aortic interspace
is weak, owing to the obstruction and diminished
amount of blood thrown into the aorta. The first
sound at the apex may be strong. The radial pulse is
regular, as a rule, but in marked cases it is small,
hard, and rigid. The physical signs are:

Inspection.—The apex-beat is displaced downward
and outward from its normal locality, in proportion to
enlargement of the left ventricle. The cardiac impulse
is usually seen to be more forcible than normal.

Palpation.—The apex-beat is felt to be somewhat
displaced downward and outward, and to be increased
in force. Basic systolic thrill is sometimes felt, but

only in those cases where dilated hypertrophy is well marked.

Percussion.—The area of dullness is enlarged, especially the area of superficial dullness, and the quality is more markedly dull, of course, in proportion to the amount of cardiac enlargement and displacement of pulmonary tissue.

Auscultation.—The first sound is usually normal or louder, but the second sound over the aortic interspace

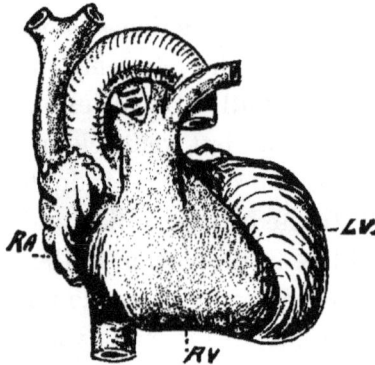

FIG. 27.—Aortic Obstruction and Regurgitation, showing Enlargement of Left Ventricle.

is weak, for reasons already given. A systolic basic murmur is heard, which has various names, such as aortic direct, obstructive, stenotic, and so on. This aortic obstructive (direct, stenotic, constrictive, systolic) murmur is heard loudest, not directly over the site of the aortic valves (p. 188), but over the aortic or second interspace near the right edge of the sternum, where the most prominent part of the bulge of the aorta lies (Gray). The murmur is systolic in time, like the mitral regurgitant, and occurs with the first sound,

or rather just before the second. It may be represented
by the following diagram:

ŭṇ *ph-*   ⁺ŭṇ

|————| · |——| · · · · |

Area of Transmission.—Though heard loudest over
the aortic interspace, the murmur, if loud enough, and
is caught up by the sternum, may be transmitted along
that bone from one end of it to the other. Besides
this, it is often conveyed by the arteries into the neck,
and sometimes may even be heard behind over the
aorta on the left side of the spinal column. If it be
loud enough, as when it is musical sometimes, it may
be heard all over the chest. In rare cases it is heard
equally loudly over the pulmonary interspace, simply
because the pulmonary artery takes up the sound from
the aorta, owing to their unusual proximity. But if
the murmur is very feeble it will with difficulty be
heard, even at the aortic interspace as sometimes happens. The loudness or feebleness of the murmur will
give little or no true idea of the real extent of the
lesion. The murmur is rendered more distinct by
walking rapidly, or in case there be co-existing anæmia. In all cases of doubt as to whether a murmur be
present or not it is well to have the patient walk briskly
up and down the room several times, and then auscul-
tate immediately upon stopping. Sometimes, however,
a purely dynamic murmur may be produced by this
means.

Diagnosis of Aortic Obstruction.—There are other
systolic murmurs to be heard at the base that very
closely resemble the true aortic obstructive, just as
there are systolic murmurs at the apex to imitate the

15

true mitral regurgitant. Omitting the pericardial and
cardio-pleuritic adventitious sounds already differentiated when speaking of mitral regurgitation (p. 206),
there remain the (1) intra-arterial and (2) intra-ventricular organic murmurs, besides the inorganic, or
functional murmurs, (3) the dynamic, and (4) the car.
dio-respiratory.

(1) *Intra-arterial Murmurs.*—These are organic systolic basic murmurs, due to roughening of the lining
membrane of the ascending portion of the aorta, inflammatory vegetations, co-arctation (bending together)
of the aorta, slight constriction, sacculation or pouch-
ing of the vessel near the heart, and pressure on the
aorta near its origin from tumors or fluid in the pericardial sac. Though sufficient to produce a systolic
basic murmur, they might not materially interfere with
the outward flow of blood, so that the left ventricle is,
in such cases, not usually enlarged, whereas in aortic
obstruction, unless of very recent occurrence, it is enlarged. Where no enlargement is observable, the history of a recent attack of endocarditis would be of
service; otherwise, it might be difficult, if not impossible, to make an absolutely correct diagnosis, even by
the pulse as traced with the sphygmograph, since it is
but little altered in aortic obstruction. Intra-arterial
murmurs are invariably loudest over the site where
they are produced, as in case of aneurismal murmurs.
It might so happen that the cause of the murmur
might be situated in the pulmonary artery, but caught
up by the aorta. For that reason, it is thought better
to term them intra-arterial than intra-aortic murmurs.

(2) *Intra-ventricular Murmurs.*—The same remarks

apply here as when speaking of mitral regurgitation.
(p. 206). Instead of the aortic orifice being affected by
the endocarditis, a lesion somewhere within the ventricle (intra-ventricular) may have been produced, sufficient to give rise to a systolic murmur, heard only, or
loudest, over the aortic interspace. It is differentiated
from aortic obstruction, not by the quality, pitch, or
other properties of the murmur, but by the fact that if
the murmur be purely intra-ventricular, the left ventricle is not enlarged.

(3) *Dynamic Murmurs.*—These murmurs are caused
by perverted action of the heart, and may be of neurotic
or anæmic origin, as described when speaking of them
in connection with mitral regurgitation (p. 207). They
occur more commonly at the base than the apex, but
are always systolic in both localities. In a word, diastolic heart murmurs are organic, and all functional, or
inorganic murmurs are systolic.

These dynamic aortic systolic murmurs of neurotic
origin are distinguished here, as elsewhere, by their
being inconstant. They are observed not only among
the choreic, nervous, and hysterical, but often among
athletes, during or immediately after violent exercise.
A perfectly healthy person, especially a young girl,
after violently running up steps, will often have a temporary dynamic basic aortic systolic murmur. Sometimes it is pulmonic also. A murmur may be sometimes produced, also, with pressure of the stethoscope
with those having thin and yielding chest walls, as
among children. Such dynamic murmurs are told by
their being inconstant, and by their not being necessarily associated with hypertrophy of the left ventricle.

Dynamic aortic systolic murmurs may also be due to anæmia, as in the case of apex murmurs. They are distinguished by their not being necessarily associated with the left ventricular enlargement; by other and co-existing signs of anæmia, of which venous hum (soon to be fully described) is one of the most important; by the non-existence of other murmurs; and by their disappearing under proper treatment for anæmia.

(4) *Cardio-respiratory murmurs*, so-called, are thrown out by having the patient hold the breath, and they then cease at once, as already described when speaking of mitral regurgitation (which see). The aortic regurgitant murmur, having a different area of transmission, and occurring in diastolic time instead of systolic, besides other signs to be described, need not be dwelt upon here. Sometimes aortic obstruction is so marked that the left ventricle will, in time, become so enlarged as to cause the mitral valves to be somewhat relatively insufficient, giving a slight mitral systolic regurgitant murmur at the apex; but this condition appears to be somewhat rare.

Anæmic murmurs are usually placed here, so that a few remarks in regard to them at once become necessary. According to some authors, they are always produced at the aortic orifice. The theory is that in anæmia the heart muscle becomes weak and flabby, and stretches so that the cavities of the heart are larger than normal, while the tough rings around the orifices remain the same. Enlarged cavities, the orifices remaining normal, is relatively the same thing as normal cavities and constricted orifices, so that such a so-called anæmic murmur is, after all, simply a sort of temporary obstruction murmur, in other words, a dynamic murmur due to anæmia, and not purely a blood, or hæmic murmur.

Purely hæmic murmurs, due to anæmia, are heard over the pulmonary interspace rather than the aortic. Sometimes, it is true, they are caught up by the aorta, just as aortic organic obstructive murmurs are sometimes caught up by the pulmonary artery when the two press closely on each other. Not only are these pulmonary hæmic murmurs sometimes caught up by the aorta, but by it they may be and are, sometimes, conveyed into the arteries of the neck. Frequently, these murmurs in the arteries of the neck, however, are not conveyed there, but are simply created by pressure from the stethoscope disturbing the regularity of the calibre of the artery.

How are basic dynamic murmurs due to anæmia to be distinguished from aortic obstructive murmurs, both being systolic? Not by their quality, pitch, or other properties, albeit murmurs dependent on anæmia for their production are usually lower in pitch, softer, more blowing in quality, and more diffused about the base of the heart, but are not transmitted so far as in the case of obstruction. But this is not always the case. The real difference is that aortic obstruction causes enlargement of the left ventricle, whereas anæmia does not, although palpitation from chronic anæmia may give rise to a somewhat general enlargement of the heart from overwork long continued. Or else the enlargement would be due to dilatation from anæmia, with feeble impulse, but in either case it would be general instead of limited to the left ventricle, as is

usually the case in aortic obstruction. Only in rare
cases of long standing does aortic obstruction give rise
to relative (secondary) mitral regurgitation with sequential enlargement of the left auricle and right ventricle. Dynamic murmurs, due to anæmia, whether
heard at the apex or base, are also associated with
other signs of anæmia, among which venous hum may
be mentioned here.

Venous hum, or *bruit de diable*, is a continuous
though remittent roaring sound, heard over certain
large veins, as the jugulars, subclavian, and femoral
veins. From the fact that the sound is continuous it
is readily distinguished from an arterial murmur. It
is most convenient to listen for it in the neck, and it is
best heard over the junction of the internal jugular
and subclavian veins. Turn the patient's head to one
side and elevate the chin, so as to render the tissues
tense on that side of the neck. Now place the stethoscope over the point mentioned, just at the root of the
neck, and the continuous roar of venous hum (not
rhythmical tracheal breathing) will be distinctly heard,
usually in an anæmic person, especially a young anæmic woman. Pressure with the finger, or otherwise,
over the vein on the distal side of the stethoscope
causes the hum to cease at once. Remove the pressure
and it immediately returns.

Guttmann, of Berlin, thinks that venous hum is due
to the vortiginous (eddying, whirling) movement of
blood in the ampullæ (bulbs) formed by the union of
the internal jugular and subclavian veins. The volume
of blood, he thinks, is smaller in anæmia, and the ampullæ remain the same size, as they are adherent to

the surrounding connective tissue. This would allow space in the ampullæ for the vortiginous or eddying movement of the blood, as mentioned. According to Guttmann, venous hum is also heard more frequently and with greater intensity on the patient's left side of the neck, since the curved direction that the blood has to take on the left side would produce more vortiginous movement. Not infrequently, however, it is heard loudest on the patient's right side of the neck, and sometimes, indeed, when it cannot be heard at all on the left. In the first place, it is highly improbable that the ampullæ would be prevented from contracting, and becoming accommodated to the diminished volume of blood, by their being adherent to surrounding and rather loose connective tissue. In the second place, venous hum does not appear to be due to the vortiginous movement of venous blood, since it is produced best where the blood has a straight and rapid course in large veins, and for that reason is heard best, or only, on the right side of the patient's neck. There are various other theories about the production of venous hum.

According to Fothergill, many ingenious hypotheses have been raised regarding the causation of venous hum, but none as yet have been accepted. According to Walshe, the composition of the blood in the anæmic state cannot be overlooked. It appears that watery venous blood is soniferous, while arterial blood, under the same conditions, is not, the exact reason for which does not appear to be known. It would seem, therefore, that venous hum depends upon the character of the venous blood in the anæmic state, rather than upon

the supposed vortiginous movement of the blood in
ampullæ, the vibration of valves of veins, and so on.

AORTIC REGURGITATION.

Aortic regurgitation (reflux, insufficiency) is the
cause of great enlargement of the left ventricle only,
as a rule. Should relative insufficiency of the mitral
valves occur, the left auricle, and, in time, possibly, the
right ventricle also, would become enlarged, though
the patient would hardly live long enough in such a
marked case. It is one of the most hopeless and
fatal of all valvular lesions of the heart. Walshe
places it fourth in the order of relative gravity, but,
leaving out tricuspid regurgitation, I should be inclined to place it first.

The reason why the left ventricle becomes enlarged
in aortic regurgitation is obvious. The blood regurgitating back into the ventricle simply gives it double
work to perform. The first sound of the heart may be
louder than normal, or it may be absent. In other
words, it varies, depending, no doubt, upon the amount
of dilatation and whether or not it has brought about
relative mitral insufficiency. The second sound over
the pulmonary interspace is usually normal, unless the
right ventricle becomes sequentially enlarged from relative insufficiency of the mitral valves, and then the
second sound over the pulmonary interspace will be
accentuated. Over the aortic interspace the second
sound is taken up by the aortic regurgitant murmur
which occurs with it, being diastolic in time. The
pulse of aortic regurgitation is characteristic. Owing
to enlargement of the left ventricle, the blood is driven

with great force into the aorta, but as that artery
empties itself in two directions at once, back into the
left ventricle and forward into the capillaries, the pulse,
which started with such a thump, suddenly collapses
from not being sustained, so that it is sometimes called
collapsing, vanishing, unsustained, locomotive, or water-
hammer pulse. There are various names by which it is
known, and we have good authorities for them, but it
would appear that water-hammer is the one now most
commonly used. The physical signs of aortic regurgitation are: (See fig. for Aortic Obstruction.)

Inspection.—The apex-beat is carried down and out,
owing to enlargement of the left ventricle. The heart's
impulse is usually seen to be more forcible than normal. There is usually pulsation in the arteries about
the neck, and all over the body where the arteries are
superficially located and visible. If the patient be directed to hold the arm up, not infrequently the radial,
ulnar, and other arteries of the upper extremity are
seen to pulsate as in arterial sclerosis.

Palpation.—The apex of the heart is felt displaced
downward and outward, and sometimes basic diastolic

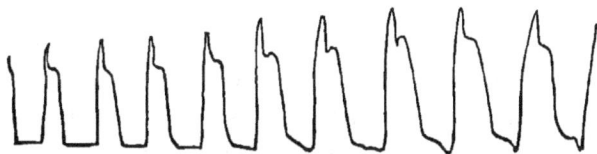

Fig. 28.—Sphygmographic Tracing of Aortic Regurgitation Pulse.

thrill is felt. The radial pulse is unsustained, or collapses under the finger. The sphygmographic tracings
of this impulse are characteristic.

Percussion.—This shows an increased area of cardiac

dullness, due to enlargement of the left ventricle. It may be still more increased if there be relative mitral insufficiency and consequent enlargement of the left auricle and also right ventricle.

Auscultation.—This reveals the presence of the characteristic aortic regurgitant murmur. It is diastolic in time, and occurs with, or takes the place of, the second sound of the heart. It is sometimes termed the aortic indirect or insufficient murmur. The aortic regurgitant (indirect, insufficient, diastolic) murmur may be represented by the following diagram:

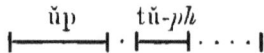

$$\text{ŭp} \qquad \text{tŭ-}ph$$
$$\vdash\!\!-\!\!-\!\!-\!\!-\!\!\dashv \cdot \vdash\!\!-\!\!-\!\!\dashv \cdot \cdot \cdot \cdot \dashv$$

The quality of the murmur varies, but in most cases it appears to be harsher and higher-pitched than the aortic obstructive murmur. Sometimes they exist together, forming what may be termed steam-tug murmur, *hoo—chee.* The combination is usually best heard on the sternum about the fourth or fifth cartilage.

Area of Transmission.—The aortic regurgitant murmur is directed back from the current of blood, and for that reason is not usually heard so plainly in the aortic interspace as it is over the sternum about the fourth cartilage. It is sometimes heard only at the lower end of the sternum. Sometimes it is conveyed backward to the apex of the heart, where it is to be distinguished by its diastolic time, diminished intensity, and the enlarged left ventricle, from mitral obstructive murmur (which see). If the murmur is loud enough, the sternum will catch up the sound, so that it may be heard from one end of that bone to the other. Occasionally, the pulmonary artery takes up the murmur from its proximity to the aorta, so that pulmonary regurgitation may be imitated (p. 239). Sometimes it is heard behind along the spinal column, but it may be loud and musical, so as to be heard as soon as one enters the room, and may be so audible to the patient as to prevent sleep.

Diagnosis of Aortic Regurgitation.—This is based upon the characteristic unsustained (water-hammer, collapsing, locomotive) aortic regurgitant pulse, enlargement of the left ventricle, and the accompanying diastolic regurgitant murmur. The latter, in rare cases, may be imitated by a murmur caused by roughness with dilatation of the ascending portion of the aorta. The late Dr. Austin Flint speaks of this as an aortic diastolic non-regurgitant, or prediastolic, murmur. It is usually preceded by an aortic direct murmur, but not always. This imitative murmur is not necessarily associated with enlargement of the left ventricle, and the pulse characteristic of aortic regurgitation is wanting. These diastolic murmurs may, in rare cases, be heard only at the lower end of the sternum, like tricuspid regurgitation, but the latter being systolic in time, the diagnosis is easy.

AORTIC OBSTRUCTION AND REGURGITATION.

This combination of lesions is second in the order of frequency, according to Walshe. Separately or together, they cause enlargement of the left ventricle. When both lesions exist at the same time, the left ventricle is enlarged and there are two murmurs, the systolic obstructive and the diastolic regurgitant already described. The two are often heard together, especially on the sternum, usually about the fourth or fifth cartilage, and form what may be termed the steam-tug murmur, *hoo—chee*, hoo standing for the obstructive, and chee for the regurgitant murmur.

PULMONARY MURMURS.

We listen over the apex, ensiform cartilage, and aortic interspace, respectively, for mitral, tricuspid, and aortic murmurs in their maximum intensity. We now come to the fourth and last point on the front of the chest walls, the pulmonary interspace, where we listen for pulmonary murmurs in their maximum intensity. The pulmonary interspace is the second interspace, or the interspace between the second and third costal cartilages, at the left edge of the sternum. Theoretically, we would have the same number of murmurs here as over the aortic interspace, with the addition of the purely anæmic or hæmic murmur. But pulmonic diastolic regurgitant murmur is so rare that it may be practically thrown out altogether. Of the systolic pulmonary murmurs the same remarks would apply here to intra-ventricular, intra-arterial, dynamic, and cardio-respiratory murmurs, as when describing aortic systolic murmurs. There remain two pulmonary murmurs, therefore, the pulmonary obstructive, and the anæmic.

PULMONARY OBSTRUCTION.

Pulmonary obstruction (constriction, stenosis) is not commonly observed among adults. It is due to fœtal endocarditis in which the right, instead of the left side of the heart is affected, as already stated (p. 193). Children are therefore born with pulmonary valvular lesions, when they have them, and usually die early. It gives rise to a basic systolic murmur, as in the case of aortic obstruction, but heard on the left side of the sternum over the pulmonary (second) interspace in its maximum intensity, instead of on the right. The pulse is not affected by it. It leads to more or less enlargement of the right ventricle. The diagram representing this murmur is the same as for aortic obstruction.

The area of transmission is the same as for the aortic obstructive murmur so far as the sternum is concerned, but beyond that it differs materially. It is not conveyed up into arteries of the neck, but back in the lungs by the pulmonary artery, and for that reason, when loud, is heard behind over both sides of the backs of children having thin chest walls. It is also transmitted out toward the left shoulder sometimes. In a case of a girl of fourteen now under my observation, the murmur is so loud as to be heard distinctly all over the chest, though loudest and most distinct over the pulmonary interspace. Systolic basic thrill is sometimes felt, as in aortic obstruction.

How are we to tell it from the anæmic or other murmur? Simply because it is usually associated with more or less cyanosis or blue disease (morbus cæruleus), owing to the venous congestion over the body. Walshe says that true pulmonary obstructive murmur is very rare, except in cases of cyanosis. Fits of dyspnœa are likely to occur any time, owing to the lungs not being properly supplied with blood. Cyanosis itself does not give rise to a murmur as anæmia does.

True anæmic or hæmic murmurs due to a watery condition of the blood are always systolic, and perhaps

always basic. They are also apparently pulmonic instead of aortic, although they may sometimes be taken
up by the aorta when the two vessels are in close contact with each other. May not these murmurs be
nothing more than venous hum in the pulmonary
artery, thrown into rhythm by proximity to the heart?
It is true that they do not always co-exist with venous
hum in the neck, but that may be because the venous
blood travels differently in those vessels.

Another supposed cause of systolic anæmic murmurs
heard over the pulmonary interspace would not be
mentioned here but for the fact that Balfour's name is
connected with it, although, according to Flint, it did
not originate with Balfour, but with Naunyn. The
latter thought that the murmur was not to be attributed to the pulmonary orifice at all, but was due to a
slight mitral regurgitation due to weakness of the
heart from anæmia, with an accompanying mitral
systolic anæmic murmur, not loud enough to be
heard at the apex but loud enough to be taken up by
the left auricular appendix, which had also become
dilated and enlarged from anæmia! Yet in mitral
regurgitation with a loud murmur and the left auricle
very much more enlarged than in anæmia, the transmission of the murmur to this point is very rare, as
already stated (see Mitral Regurgitation). Flint, perhaps justly, characterizes such reasoning as strained.
Vierordt, of Leipsic, states that the explanation of
these cardiac anæmic murmurs is very difficult, and
thinks that in many cases Sahle's suggestion might be
available, that these murmurs may arise from the large
vessels concealed in the thorax.

PULMONARY REGURGITATION.

Pulmonary regurgitation (reflux, insufficiency) is such a rare disease that it is hardly worth mentioning. If it occurred it would give rise to a pulmonary regurgitant murmur, which would be diastolic in time, and could not be confounded with any functional or inorganic murmur. As Flint states, a diagnosis could be made when other signs went to show the existence of pulmonary and the absence of aortic regurgitation. Diastolic thrill might be felt in both. But the pulse, of course, would be different, being unaffected by pulmonary regurgitation, and collapsing in the aortic lesion. Jugular pulsation would occur in time, with enlargement of the right ventricle in pulmonary regurgitation, together with cyanosis and cardiac dropsy, none of which are characteristic of aortic regurgitation. As the disease is so rare, however, it is perhaps really more of a clinical curiosity than of any practical value.

RELATIVE GRAVITY OF VALVULAR LESIONS.

According to Walshe, the following is the order of relative gravity of valvular lesions: (1) tricuspid regurgitation, (2) mitral regurgitation, (3) mitral obstruction, (4) aortic regurgitation, (5) pulmonary obstruction, (6) aortic obstruction. Very little is known about pulmonary regurgitation and tricuspid obstruction.

Neither of the first three mentioned produce what is generally understood as sudden death from heart disease. But all three are dangerous, from such complications as pulmonary congestion, hemorrhagic infarction

(p. 101), cardiac dropsy, and pulmonary œdema, which are likely to occur. But in aortic regurgitation not only is thoracic aneurism likely to be produced from the tremendous force with which the blood is thrown into the aorta (p. 271), and sudden death from cerebral apoplexy, but there is also liability to sudden death due to failure of the heart's action from some cause, supposed to be, by some, failure of blood to be conveyed by the nutrient arteries of the heart. For these reasons it would appear that aortic regurgitation, well-marked, is among the most dangerous of all valvular lesions of the heart.

ENDOCARDITIS.

Endocarditis, as already said, occurs most frequently in the course of acute articular rheumatism, especially among the young, but it may also occur in the course of any other disease, or independently. During fœtal life it attacks the right side of the heart, but after birth it attacks the left side, as already stated (p. 193). It usually results in valvular disease of some kind (mitral regurgitation, most frequently with consequent enlargement), or else some lesion elsewhere within the ventricle, or finally it may possibly end in complete recovery.

The physical signs of endocarditis are based chiefly on auscultation. Inspection usually is negative. Palpation is usually also negative, but may reveal the fact that the heart's action and pulse are excited, and sometimes irregular. If it be the first attack, percussion dullness will be very slightly, or not at all, increased in extent, as the heart will not be appreciably enlarged. Upon auscultation the sounds will be normal or perhaps slightly increased in intensity, except at the site of the murmur, which will be heard at the apex or over the aortic interspace. The murmur occurs early in the disease, and is systolic in time and blowing in quality. It is due to roughening of the surface or deposit of lymph on or near the valves of the mitral or aortic orifices. The positive proof, according to Flint, is the presence of a mitral systolic murmur occurring during an attack of rheumatism If it be aortic it is in many cases inorganic. The mitral systolic murmur need not show actual regurgitation as yet, or it may be purely intra-ventricular. But if the left ventricle be enlarged and endocarditis be suspected, it will not be the first attack, since the enlargement will be due to mitral regurgitation from a previous endocarditis.

PERICARDITIS.

Pericarditis usually occurs in connection with acute articular rheumatism, tubular nephritis, pleurisy, syphilis, pneumonia, tubercle, or cancer. Of course it may be due to surgical injury also, but it is rarely, if ever, an idiopathic affection. It is usually attended with more or less effusion, and is divided into three stages. In the first stage (congestion) the action of the heart is irritable and forcible. The area of dullness is as yet unchanged. Auscultation uniformly reveals the presence of a pericardial (exocardial) friction sound. Endocardial murmurs may also exist. But pericardial (exocardial) friction murmurs are superficial, rubbing, churning, clicking, or creaking, never blowing, whistling, or roaring, and are limited to the cardiac region, and often vary in intensity with position of the patient, or pressure with ear or stethoscope, and occur inde-

16

pendently of the heart sounds—that is, not fixed.
Endocardial murmurs, on the contrary, are fixed in
their time of occurrence with regard to the heart
sounds, and are often conveyed or transmitted over cer-

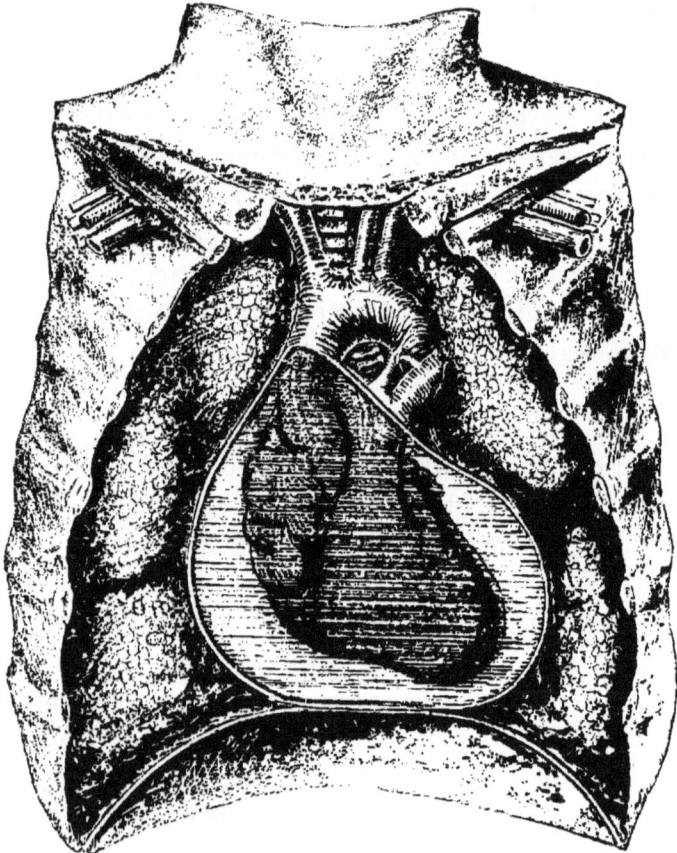

Fig. 29.—Diagram, Pericarditis with Effusion.

tain areas, as already described. The second stage of
pericarditis (effusion) is characterized by the effusion
of liquid—acute hydro-pericarditis.

Inspection now shows prominence of the præcordial

region in proportion to the amount of effusion, diminution, or absence, of the apex-beat, and diminution of the respiratory movements on the left side.

Palpation.—The apex-beat is raised upward and outward to the left, is feeble or suppressed, and may change with position of the patient. If not felt when the patient is on the back, it may be perceptible if the patient leans forward. An undulating impulse is sometimes felt, and the epigastrium may be bulging from depression of the diaphragm.

Percussion.—The area of præcordial dullness is enlarged. If the effusion be great this area is found to be wider below than above, on account of the shape of the pericardial sac. It may extend down to the seventh rib and up to the first rib, and from nipple to nipple, or even further.

Auscultation.—The pericardial friction sounds have now disappeared, since the two surfaces cannot rub together. The heart sounds are feeble and heard better at the top of the sternum than elsewhere, as the effusion occupies less space there than below. Sometimes a basic systolic murmur, due to pressure of the effusion on the aorta, is heard. The respiratory murmur, pectorophony (vocal resonance), and vocal fremitus are diminished or absent over the central portion of the cardiac region. The third stage is that of absorption. The friction sound returns—*frictio redux*—the heart sounds become more distinct, and there is a gradual return to health in favorable cases. In other cases it may become subacute or chronic.

Chronic pericarditis may be attended with adhesions simply, or there may be adhesions with hypertrophy or atrophy (which see), or there may be chronic pericarditis with effusion, termed chronic hydro-pericarditis. The physical signs of chronic hydro-pericarditis are similar to those of the acute variety just described.

Pneumo-hydro-pericarditis, air and fluid in the pericardium, is rare, but may possibly result from decomposition of liquid effusion, or, usually, from perforation communicating with the œsophagus or lungs. In these cases the heart's action is accompanied by metallic (amphoric) tinkling, amphoric (metallic) buzzing, and splashing like a water-wheel (Walshe).

Hæmo-pericardium, or blood in the pericardium, may result from cancer, scurvy, and the hemorrhagic tendency, or surgical injury, and rupture of aneurism. If the patient lived long enough the blood, while fluid, would give the same signs as hydro-pericarditis.

Hydro-pericardium, or dropsy of the pericardium, is merely associated with general dropsy from Bright's disease of the kidneys, or some other cause, without inflammation of the pericardium. The physical signs are those of hydro-pericarditis, already described.

Pneumo-pericardium, or air in the pericardium, is due to gas arising from *post-mortem* changes, as a rule, and is rarely, if ever, seen during life, and *pneumo-hydro-pericardium*, if such a thing could exist during life, would give the same physical signs as pneumo-hydro-pericarditis, already described.

MYOCARDITIS.

Myocarditis, or carditis, signifies inflammation of the heart muscle itself. When present it is usually associated with endocarditis or, more frequently, with pericarditis. In the acute form, the pain in the cardiac region is extreme, the pulse rapid and weak. In the chronic form, the symptoms are those of a weak heart, with want of correspondence between the heart and pulse beat. Neither form is diagnosticated during life.

HYPERTROPHY OF THE HEART.

Hypertrophy of the heart differs from hyperplasia. In the former case the existing anatomical elements are enlarged, in the latter their number is increased (Flint). In both cases the heart is enlarged.

Hypertrophy may be general, or limited to one or more of its compartments. In the latter case the left ventricle is by far the most frequently affected, then the left auricle, then the right ventricle, and lastly the right auricle may sometimes be somewhat enlarged. Again, the hypertrophy may be concentric, simple, or eccentric. In the first case the walls are thickened and the cavities become smaller. This is so rare that it is more theoretical than practical, and may be thrown out altogether. Simple enlargement is not common either —that is, thickening of the walls—the cavities remaining the same. Indeed, hypertrophy or enlargement of the heart from any cause is so uniformly of the eccentric variety, that, unless otherwise specified, enlargement of the heart will always be meant to be eccentric hypertrophy—that is, hypertrophy with dilatation, usually termed hypertrophous dilatation, or dilated hypertrophy.

Of the causes of enlargement of the heart, (1)*valvular lesions*, as previously stated, are the most frequent.

Moreover, each of these lesions gives rise to its own characteristic enlargement, which will be marked in proportion to the degree of the lesions. Thus, mitral regurgitation causes enlargement (dilated hypertrophy) of the left auricle, left ventricle, and right ventricle in the order named. Mitral obstruction causes enlarged left auricle and right ventricle. Aortic obstruction gives rise to enlarged left ventricle and aortic regurgitation also, only the enlargement is usually more marked in aortic regurgitation; and in either case, but especially in aortic regurgitation, there may occur relative insufficiency of the mitral valves, with consequent enlargement of the left auricle, if not the right ventricle, also, in time. Pulmonary (pulmonic) obstruction or regurgitation will lead to enlargement of the right ventricle as would also tricuspid regurgitation, relative, or resulting from fœtal endocarditis, while tricuspid obstruction would cause enlargement of the right auricle only.

(2) *Bright's Disease of the Kidneys.*—In the chronic interstitial variety there is enlargement of the left ventricle only, and this is not due to valvular lesion but simply to the fact that the left ventricle has extra work to perform in overcoming the obstruction due to the lessened calibre (*lumen*) of the arterioles throughout the body. In the chronic tubular variety of Bright's disease of the kidneys, there is liability to inflammation of serous membranes generally, the endocardium included, with resulting valvular lesion. Enlargement of the heart, therefore, in chronic tubular nephritis is usually due to valvular lesion, resulting from an endocarditis.

(3) *General Vesicular Emphysema.*—In this case there is enlargement of the right ventricle, owing to obstruction to the pulmonary circulation.

(4) *Exophthalmic goitre,* called also cardio-thyroid exophthalmos, Basedow's, or Graves' disease. Here we have general cardiac enlargement due to over nourishment from vaso-motor dilatation of the nutrient vessels of the heart. According to Niemeyer, Bamberger, and others, the nutrient vessels of the heart in this disease are enlarged from vaso-motor dilatation due to some disturbance of the cervical ganglia of the sympathetic nervous system (see Exophthalmic Goitre).

(5) *Palpitation* from anæmia, or other cause, may give rise to enlargement of the heart from its overwork. The alcoholic habit probably acts somewhat in this way. Habit, mode of life, occupation requiring prolonged muscular exertion, as among athletes, and the like, also enlarge the heart to a certain extent, as well as excessive venery. The physical signs of hypertrophy of the heart with dilatation are usually more marked than they are when simple hypertrophy alone exists. They are:

Inspection.—The enlargement is always more to the patient's left than right, and the line of the base is rarely, if ever, changed. The extent of the visible impulse is increased, and there is more or less prominence of the præcordial region. The apex-beat is also seen to be more forcible than normal, and it may be as low as the ninth rib and outside the mammillary line. Enlargement of the right ventricle pushes the apex further to the left than normal, but the apex is also lowered when the left ventricle is enlarged.

Palpation.—The impulse is heaving and lifting in character, with or without thrill, and its area is increased. Hypertrophy of the right ventricle usually gives a strong epigastric impulse. When the left ventricle is hypertrophied the apex-beat is carried down and out. Præsystolic impulse is sometimes felt over an hypertrophied left auricle, as may occur in mitral disease.

The radial pulse in hypertrophy of the right side of the heart is not appreciably affected in character. But in hypertrophy of the left ventricle without regurgitation or obstruction, the radial pulse is full, prolonged, and sustained.

Percussion.—Both areas of dullness are increased, laterally and vertically. General enlargement may give dullness on percussion from the third to the eighth rib, and from an inch to the right of the sternum to two or three inches outside the left nipple. Walshe mentions a case where enlargement (dilated hypertrophy) of the heart was so extensive as to be mistaken for pleurisy of the left side with effusion. Hypertrophy of the left ventricle gives dullness usually beyond the left nipple; of the right ventricle, considerably to the right of the sternum. In hypertrophy of the left auricle the area of dullness over that portion is enlarged and more marked.

Auscultation.—The first sound, dull, muffled, prolonged, diffused over a larger area than in health, and increased in intensity, may indeed so closely resemble a slight systolic murmur as to make it sometimes difficult to decide; and the second sound is also louder and more diffused than in health. If murmurs are present they will more or less obscure, or take the place of the heart sounds. There is diminution or absence of the respiratory murmur over the præcordial space.

Hypertrophy of the right auricle rarely occurs, and then it is due to tricuspid regurgitation usually, as tricuspid obstruction (stenosis, constriction) is almost unknown during life (Walshe).

DILATATION OF THE HEART.

Dilatation of the heart may be one of three kinds: (1) hypertrophous dilatation, or dilated hypertrophy, which is the most common form, and just considered; (2) simple dilatation, where the walls remain the same but the cavities are enlarged; and (3) attenuated dilatation (Walshe), where the cavities are not only enlarged, but the walls are thinner than normal. It is the last variety that requires our attention. Attenuated dilatation, or *dilatation without compensating hypertrophy*, is a hopeless disease. It may result from valvular lesion, or general vesicular emphysema, where dilatation and hypertrophy of certain parts occur together, producing the various enlargements characteristic of those diseases, as already fully described. Presently the time arrives, however, when hypertrophy ceases to compensate, and then the case becomes one of dilatation, since, evidently, enlargement cannot continue to go on indefinitely. In other cases dilatation occurs from inherent weakness of the heart muscle itself, in other words, from no known cause. Thoracic aneurism causes dilatation in so far as it is a cause of obstruction to the outflow of blood, and in this way acts as valvular lesion would. But the underlying cause of aneurism, if it be gout, syphilis, or lead poisoning, may cause weakness of the heart muscle also, and thus favor dilatation. The physical signs of cardiac dilatation are:

Inspection.—The visible area of the apex-beat, if indeed it be visible, is increased without any particular point of maximum intensity. Dyspnœa and cyanosis are sometimes observed, especially after attempted exertion.

Palpation.—The cardiac impulse is feeble, its area is increased, rarely is there any thrill, but rather an undulating motion over the præcordial region, especially if there be mitral regurgitation. The radial pulse is feeble, sometimes irregular, small, and compressible.

Percussion.—The area of cardiac dullness is increased in the direction of the part dilated, or generally increased if the dilatation affect the whole heart. In the latter case it is oval or somewhat square in shape, instead of being triangular, with the base downward, as in pericarditis with effusion.

Auscultation.—Both sounds are short, abrupt, feeble, and equal in duration, the second being often inaudible at the apex. The post-systolic or first period of silence is prolonged. Endocardial murmurs, when present, are indistinct. The respiratory murmur is diminished or absent over the præcordial region, owing to the cardiac enlargement.

FATTY DEGENERATION OF THE HEART.

Fatty heart is of two kinds. (1) that in which the fat is added to the organ without or within, or between its fibres, and causing trouble by pressure; and (2) that in which the muscular fibre is replaced by fatty tissue. The first is by Walshe termed fatty infiltration; and is simply an accumulation of fat. The second is known as Quain's fatty degeneration, and by Walshe is termed fatty metamorphosis of the heart. The first variety may give some inconvenience, but may be modified, if not entirely got rid of, by a Carlsbad course, or restricted diet, if thought necessary. The second is a serious and often fatal disease. There is no known cause for Quain's fatty heart. Obesity, though associated with fatty infiltration, bears no relation whatever to fatty metamorphosis. Cardiac fatty metamorphosis occurs at middle life, or past, and in men rather than women. It also occurs more frequently among the upper classes than among laborers. It is probable that any condition that interferes with the proper nutrition of the heart, such as sclerosis of its nutrient vessels from various causes, and leading to cardiac ischæmia (see Angina Pectoris), may predispose to fatty metamorphosis. But in tuberculous and other wasting diseases the heart is more frequently normal than fatty, and hence they cannot be said to be causes of the disease. The real cause, whatever it may be, is probably inherited rather than acquired. The physical signs of fatty metamorphosis of the heart (Quain's) are as follows:

Inspection.—The heart's impulse is usually not observable, owing to its feebleness. The patient may be observed to be suffering with a fit of dyspnœa and having a peculiarly anxious expression. The *arcus senilis* may be present, but appears to bear no fixed relation to the disease.

Palpation.—The impulse is so weak as to be scarcely felt, even though the patient be emaciated and leans forward. If the heart was hypertrophied first, there may be an undulating impulse, as in attenuated dilatation. The pulse is feeble and sometimes abnormally slow; or it may be irregular and intermitting, changing from abnormal slowness to rapidity—from 20 to 30 beats per minute to 150, but always weak.

Percussion.—The area of dullness will be normal, unless hypertrophy co-exists, when it would be larger, or smaller if there be atrophy.

Auscultation.—The first sound of the heart, even at the apex, instead of being somewhat low pitched and of well-marked duration, as in health, now becomes short, high pitched, and weak, and the first rest is notably prolonged. The second sound is feeble but distinct, and is accentuated in the aortic or pulmonary interspace, according as the right or left ventricle is chiefly affected. Of course, murmurs of various kinds may be present, but they are rare, and when they are present they are usually weak. The patient feels better, according to the late Dr. Alonzo Clark, lying down with the head low, as the heart would then have less to do.

ATROPHY OF THE HEART.

Atrophy of the whole heart, unless it be senile, is of rare occurrence. It sometimes takes place in connection with wasting diseases like phthisis, suppurating bone, calcification of the coronary arteries, tightly adherent pericardium, and, rarely, after pregnancy. Local atrophy of some part of the heart is more common, and

occurs in connection with fatty heart. In mitral obstruction, also, the left ventricle is somewhat atrophied,
and when this lesion occurs in children, there appears
to be usually more or less deformity of the chest, resembling the so-called pigeon breast. The flattening
is particularly well marked in the lower præcordial
region to the left of the sternum, but whether it be
due to atrophy of the left ventricle, or a lack of general nutrition from imperfect cardiac function, is not
exactly known. In case of general atrophy, the area
of percussion dullness is diminished, the heart's sounds
are clear, the impulse is feeble, the pulse quick and
feeble, but regular, and there is a great tendency to
palpitation (Da Costa).

CARDIAC DROPSY.

Cardiac dropsy usually begins about the feet and
ankles, and, gradually extending upward, is afterward
met with in various localities. It is most constantly
associated with dilatation of the right heart, as in tricuspid regurgitation, but there are exceptions. Albumen, when present in the urine, is due to renal congestion simply, unless there be also co-existing structural
lesion of the kidneys.

EXOPHTHALMIC GOITRE.

Exophthalmic goitre (cardio-thyroid exophthalmos,
Basedow's disease, Grave's disease), in order to be complete, consists of three factors: (1) palpitation and enlargement of the heart, (2) enlargement of the thyroid
gland, with throbbing of the arteries about the neck,
and (3) protrusion of the eyeballs. The disease is said

to be due to some change in, or pressure on, the cervical ganglia of the sympathetic system of nerves, which
send branches directly, or indirectly, to the three localities mentioned.

The disease is usually classed among the neuroses, or
functional disturbances, of the heart, but inasmuch as
it leads to cardiac hypertrophy, I have thought it best
to place it among the organic diseases of the heart. It
occurs more frequently among women than men, and
in most cases the women are usually nervous, if not
hysterical, and anæmic. Sometimes it occurs in men,
and Graefe mentions the case of a young man in whom
it was suddenly developed on account of nervousness
at the prospect of being married. (!) In most cases it
develops slowly, but in some instances (as in the case of
the young man just cited) it may develop very suddenly.

(1) Palpitation of the heart usually first attracts the
attention of the patient. The heart is observed to beat
120 or even 140 times to the minute, instead of 60 to
70. Enlargement of the heart follows, partly from
palpitation, but chiefly from overnutrition of the organ.
The nutrient vessels of the heart become dilated from
vaso-motor disturbances, and the heart receives more
blood supply than normal. Hence its palpitation as
well as overgrowth.

(2) The thyroid gland now begins to enlarge, and
throbbing of the inferior thyroid, carotid, and, sometimes, temporal arteries is observed.

The enlargement of the thyroid gland may not be
very perceptible, and when present, is due to dilatation
of its vessels, serous infiltration, and hyperplasia of its

tissues. The gland is rarely so much enlarged as in simple goitre (cretinism, Derbyshire neck), where it may be enormously hypertrophied or increased in size by fibrous or calcareous deposits.

Cretinism (with goitre), so-called from deformity

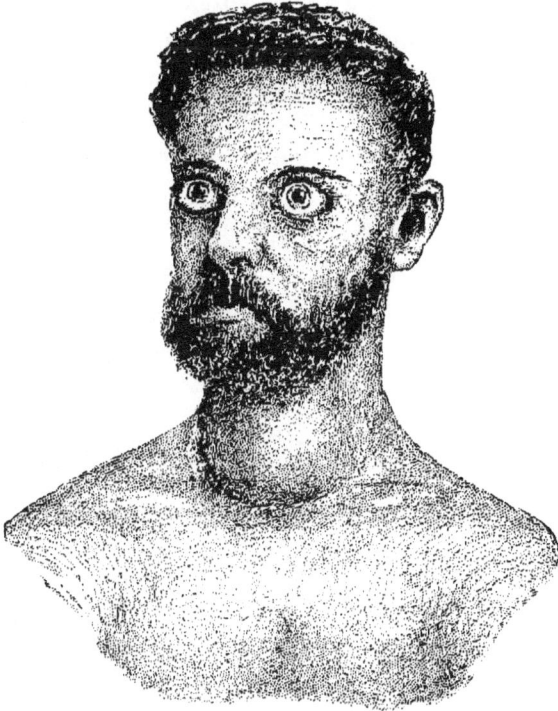

Fig. 30.—Exophthalmic Goitre. (After Eichorst.)

or mutilation, supposed to result from intermarriage, is a local disease, and found not only among those who intermarry, but also among young girls and women who carry heavy burdens and are habitually subjected to bad hygienic conditions, as in certain parts of the old world. For these reasons it is thought that the thyroid gland becomes enlarged, as in foetal life, because the lungs are not equal to the task put upon them of aërating the blood, and not because of any particular lack of iodine in the water, which is the same as it was in those localities a thousand years ago. Hence cretinism (with goitre) is becoming less, thanks to the steam-engine, which enables people to leave home and marry elsewhere, and other advances made by Christian civilization, and not because of any change in the drinking water of certain localities. To return to exophthalmic goitre, which disease is not confined to any locality, but which may, and does, occur in all parts of the world. We have seen why the heart and thyroid gland are enlarged—on account of the vaso-motor dilatation of their blood-vessels. The arteries about the neck and temples sometimes throb because they become dilated, and the blood is sent through them with great force by the hypertrophied heart. (3) Lastly, the eyes in this disease protrude, because of the increase of the fat at the bottom of the orbit. The intra-orbital fat is increased from hyperplasia, and, according to Niemeyer, it may become not only hyperæmic, but also œdematous. Sometimes the eyes protrude so that the lids cannot be closed, and consequently ulceration of the cornea, from foreign particles, may result. The upper lid, according to Graefe, becomes fixed early in the disease, from spasm of the levator muscle, and consequently does not follow the eye in looking downward.

All three factors of this disease are not equally present in every case. The heart may be chiefly affected, with scarcely perceptible change in the thyroid gland, while the eyes remain perfectly normal, and so on. In course of time, however, all the phenomena are apt to appear. Anæmia is observable in most cases before treatment, and marked venous hum, with or without thrill, in the neck is pretty constant. Not infrequently a true anæmic (hæmic) murmur is heard with systole over the pulmonary interspace; and over the aortic interspace a loud, systolic, dynamic murmur, owing to the force, I presume, with which the arterial blood is forced at times through the aortic orifice by the hypertrophied left ventricle; but, unlike the anæmic murmur over the pulmonic (pulmonary) interspace, it is inconstant, like all other dynamic murmurs. The physical signs of cardiac hypertrophy have already been considered (p. 247). Unless the patient dies from apoplexy or rarely, suffocation due to pressure on the trachea, the prognosis is not bad. Recovery, according to Niemeyer, is more common in this disease than death.

ANGINA PECTORIS.

Angina pectoris, or suffocative breast-pang, is usually defined to be a paroxysmal neurosis of the heart, and always attended with pain, whatever other symptoms may be present. As it is usually accompanied or preceded by organic changes in the heart, I have thought it best to place it among the organic, rather than functional, diseases of that organ.

The disease is of two kinds, true and false. The latter is almost wholly confined to young hysterical women. In these cases the pain does not extend through to the back and up to the neck and down the left arm, but is simply located apparently in the chest

17

wall, like an intercostal neuralgia, but attended with palpitation of the heart and dyspnœa.

True angina pectoris, on the other hand, attacks men usually, and those in the upper walks of life, either at middle age or past. It depends upon what is termed ischæmia of the myocardium, which is simply a local anæmia of the heart, due to periarteritis or sclerosis of its nutrient vessels, the latter resulting from endarteritis, and often terminating in ossification. Pain is not always present in all cases of cardiac arterial sclerosis, for this condition gives rise to different varieties of symptoms, as we shall see; but when pain is present, it is analogous to the pain observed in senile gangrene due to arterial obstruction. The causes of localized anæmia or ischæmia of the heart are those which produce sclerosis of its nutrient vessels. These are, according to Huchard, of Paris, (1) toxic, as alcohol, tobacco, especially cigarette smoking, malaria, and lead, (2) diathetic, as gout, rheumatism, and syphilis, and (3) physical, moral, and intellectual overpressure.

Obliterating arteritis of the small coronary vessels of the heart is the lesion commonly found, and if this arteritis be slow, time for compensating hypertrophy may be allowed. But if the arteritis is rapid in its progress, dilatation from weakness of the heart's walls, or fatty degeneration, results. Tobacco, it would appear, is more destructive to the heart than alcohol.

Cardiac arterial sclerosis gives rise to five different forms of symptoms: (1) the pulmonary form with symptoms of cardiac asthma, so-called (see Asthma); (2) the painful form of true angina pectoris; (3) the tachycardiac form, in which there is simply rapidity

and weakness of the heart's action; (4) the arythmic
form, in which the rhythm becomes irregular, as often
occurs from the use of tobacco and on account of dyspepsia; and (5) the asystolic form, in which rapid dilatation occurs, due to weakening of the walls of the
heart from the nutrient arterial sclerosis. Of these five
forms of manifestations of cardiac arterial sclerosis,
and consequent ischæmia of the heart, only one, the
painful form, or angina pectoris proper, will be considered here. A patient, however, who has this disease
manifested in one form, may have a return of it in any
of the other four forms, for the same form does not
necessarily return every time, when once commenced.

Angina pectoris usually comes on without warning.
The patient may be asleep in bed, or it may be after a
heavy meal, or during a fit of anger, or while walking
briskly, especially up hill, against a stiff breeze. Suddenly a pain, like a death pang, pierces him through
and through in the lower præcordial region. It not
only extends through to the back, but up to the neck
and down the left arm, usually, but sometimes both
arms, and even one or both of the lower extremities.
This fearful neuralgic pain is thought to originate in
the cardiac plexus of nerves, and extends not only to
the parts mentioned, but along the gastric branches of
the pneumogastric nerve, as evidenced by the belching
of wind and sometimes even vomiting. The pulmonary branches of the pneumogastric nerve, on the other
hand, seem to escape in this painful form, since the
patient not only can breathe freely, but sometimes a
deep breath will give speedy relief. The attack may
last from a few seconds or minutes to an hour or more,

if the patient lives, and may consist of one prolonged
attack or many separate and distinct attacks. In the
latter case it may last several days or more. During
the attack the pulse may not be notably affected, but
if the attack be prolonged the pulse becomes more frequent and feeble, and the patient may die of sheer exhaustion, or else suddenly from paralysis, and not
spasm, of the heart. In either case it is a most cruel
death.

A patient who has had angina pectoris will surely
have a return of it in time, unless the case be not far
advanced, and the utmost care is observed. Death may
occur in the first attack, or in any of those succeeding.
It is absolutely uncertain. Arnold, of Rugby, died
with it, as did the great John Hunter.

In a lesser degree than just described, a patient may
only have attacks of irregular rhythm (arythmic) or
rapid pulse (tachycardia), with pain in the left arm.
This pain may extend down the whole arm, or be confined to the left wrist only, or be indefinitely situated
somewhere in the left breast. Or it may change from
one of these localities to the other, now in the left arm,
now only at the wrist, or even in one or more fingers of
the left hand, now in the left shoulder.

DISEASES OF THE HEART THAT CAUSE SUDDEN DEATH.

Sudden death from heart disease is not so common
as the laity generally suppose. There are certain forms
of disease of the heart, however, which do undoubtedly
cause sudden death, while other forms, though fatal
from the first, give rise to complications which produce

death indirectly rather than directly through the heart itself.

(1) *Aortic Regurgitation.*—Not only may sudden death from cerebral apoplexy occur in this disease, due to the force with which the blood is driven from the left ventricle, owing to the existing dilated hypertrophy of that part of the heart, but it may also occur in some way not yet thoroughly understood, but thought by some to be due to failure of the circulation in the nutrient vessels of the heart. It is claimed by some that, owing to the enormous enlargement of the left ventricle, which sometimes occurs in this disease, the coronary arteries are so pressed upon that blood cannot enter them, and the heart fails from want of blood supply. A case in point was that of a man of excellent habits, aged thirty-five, and otherwise in apparently good health. He was seen by Dr. Francis Delafield, of this city, at my request, and the diagnosis confirmed. He died suddenly on his stairway, without cerebral lesion.

(2) *Angina Pectoris.*—In this disease, as just described, the patient may die suddenly, not from spasm, but from paralysis of the heart, from failure of the coronary circulation, due to cardiac arterial sclerosis, and shock from the terrific pain.

(3) *Fatty Degeneration, or Metamorphosis (Quain's).* —This also may lead to sudden death from sudden failure of the heart to act, or from rupture.

(4) *Extreme Dilatation.*—In this disease, also, the heart may suddenly fail or rupture. Dropsy and other complications, however, may and are likely to cause death before such a sudden catastrophe.

(5) *Aneurism of the Heart.*—Usually situated in the wall of the left ventricle near the apex, and almost impossible to distinguish between it and mitral regurgitation. Both are accompanied with an apex systolic murmur, and both cause enlargement of the left ventricle. But in mitral regurgitation the murmur is usually louder than in cardiac aneurism, and more frequently heard posteriorly, besides the accompanying enlargement of the right ventricle, with accentuation of the second sound over the pulmonary interspace, which are also usually observed in mitral regurgitation but not in cardiac aneurism. Sudden death in the latter disease is generally due to rupture of the heart.

(6) *Fibrosis* of the heart, as sometimes results from syphilis, alcoholism, gout, rheumatism, and lead poisoning. The muscular tissue becomes more and more replaced by connective tissue, instead of fat, as in fatty degeneration, until finally it fails to act.

(7) *Bright's Disease.*—Lastly, in chronic interstitial nephritis, the left ventricle becomes enlarged, chiefly hypertrophied, while the arterioles are fibrosed and brittle. Hence cerebral apoplexy and sudden death not unfrequently occur in this disease. In many cases, it is preceded by retinal apoplexy, plainly to be observed with the ophthalmoscope, as in a typical case in which I called Dr. David Webster, of this city, in consultation about a year ago.

Mitral regurgitation and obstruction are both almost necessarily fatal diseases when they occur early in life. They both lead to constant pulmonary congestion, with sequential dilated hypertrophy of the right ventricle and consequent hemorrhagic pulmonary infarction, or

cardiac dropsy, or both, as has been already described.
The liver, spleen, kidneys, and gastro-intestinal tract
are subject to repeated, if not chronic, congestion, with
all the train of evils belonging to such a condition.

FUNCTIONAL DISEASES OF THE HEART.

Palpitation, irregular rhythm, pain and syncope or
fainting, are the chief so-called functional diseases of
the heart. Little need be said about fainting and neuralgia as functional cardiac diseases, since they are
found among the nervous and hysterical—chiefly anæmic or spoilt and over-petted young women. Among
the aged, or those having weak heart due to organic
change, syncope has more significance. In any case,
the patient lies down or falls, either one of which usually causes reaction, as the heart has less to do with
the body in the recumbent position. In those cases of
neuralgia of the heart occurring in men of middle life
or past, if the pain extend to the left arm and be accompanied by palpitation or irregular rhythm, it usually indicates cardiac arterial sclerosis, as already
described.

Palpitation.—All authors nearly agree that by palpitation of the heart is meant increased force of the
heart's action as well as increased frequency. Flint,
however, says that sometimes the heart's action may
be feeble. Walshe describes three kinds of palpitation: (1) simple palpitation, where the heart's force is
increased but the rhythm is regular and there is no
increase of frequency; (2) irregularity in force and
rhythm, occurring in paroxysms; and (3) increased
frequency, with diminution of force. The second vari-

ety is the one usually referred to when speaking of palpitation.

Of course palpitation of any kind may be coincident with, or due to, organic disease of the heart. But it may occur also in a perfectly normal heart, as proven by the normal size of the organ, which may be verified by the various methods of physical examination already described, the normal heart sounds, and the absence of adventitious sounds or murmurs. The causes of functional cardiac palpitation and irregular rhythm are so nearly allied that in stating one we state both.

Irregular Rhythm.—This usually occurs with irregular force also, and may be only momentary or last for several days or more. There is every conceivable kind of irregular rhythm, the enumeration of which, as a learned author remarks, would allay curiosity rather than prove useful. Sometimes the heart intermits, with, of course, corresponding intermission in the radial pulse. Sometimes the radial pulse intermits when the heart does not. This may occur in two ways. If the heart is beating frequently and feebly, the pulse wave may not reach the wrist every time, though the heart does not intermit. This is termed false intermission. Again, while the left ventricle contracts once, the right ventricle may beat twice, giving two systolic cardiac shocks, with only one radial pulse. This is termed bigemmeny. These terms could be multiplied, but are of no practical use.

Etiology.—Irregular rhythm, including intermission of the heart, may, like palpitation, be associated with, or even due to, organic cardiac disease. More frequently, however, they are both functional. There

appears to be no valvular lesion, in spite of the state-
ment of authors, that is characterized by any particular
palpitation or irregular rhythm.   Cardiac arterial sclerosis, fatty metamorphosis (Quain's), fibrosis, and aneurism of the heart, on the other hand, are productive of
paroxysms of palpitation and irregular rhythm of every
known kind.   They are to be known by the physical
and other signs of those diseases already described—
never losing sight of the pain in the left arm.   The
causes of functional palpitation and irregular rhythm
may be stated to be very much as Walshe has said.

(1) *Centric.*—Chorea, epilepsy, hysteria, and cerebral
and spinal irritation from any causes, especially cerebral
irritation attended by insomnia, as from over study.
The boy pianist, Joseph Hoffmann, recently was prevented from giving concerts in this city at the instance
of the Society for the Prevention of Cruelty to Children, one of the signs of his overwork being, as was
testified to by one of his attending physicians, intermittent pulse, which proved to be due to cerebral as
well as spinal irritation from overwork, both mental
and bodily.

(2) *Reflex, or Eccentric Causes.*—They include dyspepsia in all its forms; intestinal irritation from
worms or any other cause; articles of diet, as coffee,
tea, and alcohol with many people; genito-urinary
irritation, as seen in cases of gonorrhœa, cystitis, uterine catarrh, old stricture, ovarian diseases, and such
like, including piles, fissure of the rectum, rectitis and
proctitis.

(3) *Blood Poisoning and Anæmia.*—Tobacco undoubtedly plays a very important part in the produc-

tion of heart diseases, both functional and organic. Except it be cocaine, there is, perhaps, no more pernicious habit than the tobacco habit, especially, it is said, the smoking of cigarettes. Not all is positively known about the baneful effect of tobacco on the heart yet, but enough is known to make it certain that it causes palpitation, irregular rhythm, and cardiac arterial sclerosis with fatal angina pectoris. The mode in which this last condition is brought about from the tobacco habit is not certainly known, but it is so all the same. By tobacco habit is meant not one cigar or cigarette a day, or even two. Different people are differently affected by it. One or two cigars per day is habit for some, while others may smoke several, besides chewing. The worst cases of poisoning, and the so-called tobacco hearts, occur among tenement-house girls in this city who make cigars or cigarettes, or strip tobacco, as it is called. Especially in the winter does this occur, when they work with the windows closed and are inhaling the dust and fumes of tobacco. Irregular rhythm, palpitation, and anæmia are common among those girls. Besides tobacco, there are the poisons of opium, malaria, and also of syphilis, lead, gout, rheumatism, and Bright's disease of the kidneys.

(4) *Mechanical.*—We see this in general emphysema, where, owing to obstruction to the pulmonary circulation, the heart becomes tired, and not only becomes irregular in rhythm at times, but often intermits so as to take a rest. In pressure from tight lacing, effusions from pleurisy, in pneumothorax, ovarian and other tumors, pregnancy, aneurism, and such like causes, the heart may palpitate, or intermit. In treating these

symptoms, therefore, the necessity of first ascertaining the cause in each case, with a view to its removal, if possible, is apparent.

THE SPHYGMOGRAPH.

The sphygmograph is an instrument used in obtaining graphic representations of the pulse. In like manner the cardiograph is used with regard to the impulse of the heart. Both instruments require great care in their use, as well as experience. Much time is often consumed in endeavoring to obtain these tracings, and as they are often quite unnecessary in making a diagnosis such instruments are not likely to be of much value to the average practitioner. A few remarks regarding the sphygmograph may not, however, be out of place. There are many of these instruments in use, but perhaps Marey's or Dudgeon's is as good as any. In fact, Dudgeon's is very readily applied, more so, in my experience, than any of the rest. According to Walshe, a pulse trace consists in a series of figures representing the successive cardiac circuits, or revolutions.

Each figure consists of three parts for consideration: (1) the percussion stroke (up-stroke, line of ascent), (2) the apex, and (3) the downstroke (line of descent). The percussion, or upward stroke shows the force and character of the pulse beat during ventricular systole. The apex is broad, medium or sharply pointed, according as the pulse is more or less sustained, so that it is broad in hypertrophy, the valves being perfect, but pointed in aortic regurgitation. Just as the point of the tracer falls a little, it rises again, forming what is termed the tidal wave. The percussion-stroke (up-stroke), the apex, and the tidal wave all belong to the first sound and systole. Now comes the first period of silence, and the point of the tracer immediately drops into the aortic notch at the same time that the second sound is produced. Then follows the second period of rest, corresponding to the remainder of the downward-stroke (line of descent), marked first by the dicrotic wave, secondly, sometimes by a tricotic wave, or if there are many such waves, this part of the line of de-scent becomes polycrotic, or tremulous. Dicrotism, as well as tricotism, or even polycrotism, is due simply to the elastic recoil of the arteries, and need not be due to any abnormal condition, especially dicrotism. With the normal pulse, the percussion (up) strokes should be of the same length, so that the base, or respiration line, as it is called, of all the figures should be even and hori-zontal, as well as the apex line. In disease, however, these lines are subject to great irregularity. The sphygmographic tracings characteristic of various car-diac diseases are given at the time of describing those diseases, to which the reader is referred.

AORTIC ANEURISM.

The aorta is divided anatomically into three parts: (1) arch, (2) thoracic aorta, and (3) abdominal aorta.

The arch consists of three portions, (1) ascending, (2) transverse, and (3) descending portion. (1) The ascend-ing portion of the arch, about two inches long, arises from the upper part of the left ventricle, on a level with the lower border of the left third costal cartilage, and behind the left edge of the sternum, behind and a little below, as well as to the right of the origin of the pulmonary artery (p. 187). It passes obliquely upward and to the right, to the upper border of the right second costo-sternal articulation. A needle pushed into the second interspace on the right, close to the right edge of the sternum, would penetrate the most prominent bulge of this portion of the aorta, and hence this space, as stated before, is termed the aortic interspace. (2) The transverse portion of the arch commences at the upper border of the right second costo-sternal articulation, and arches from right to left, and from before backward, in front of the trachea and œsophagus, to the left side of the body of the third dorsal vertebra. (3) The descending portion of the arch extends from the left side of the body of the third dorsal vertebra down to the lower border of the left side of the body of the fourth dorsal vertebra.

The thoracic aorta commences at the left lower border of the fourth dorsal vertebra, and ends in front of the body of the last (twelfth) dorsal vertebra, at the aortic opening in the diaphragm, where it becomes abdominal.

The abdominal aorta commences at the aortic opening of the diaphragm, in front of the body of the last (twelfth) dorsal vertebra, and descending a little to the left side of the vertebral column, terminates on the body of the fourth lumbar vertebra, commonly a little to the left of the median line, where it divides into the two common iliac arteries (Gray).

Aneurism signifies a dilatation. According to Walshe, it is, in its widest sense, a local increase of calibre of an artery. Aortic aneurism, therefore, is a

local increase of calibre, or a dilatation of the aorta in
some part of its course. If it affect the aorta in any
part of its course within the thorax, it is termed thoracic aneurism, whether it be any portion of the arch,
or thoracic aorta. It is termed abdominal aneurism
when it affects the abdominal aorta in any part of its
course.

Classification.—There are various classifications of
aortic aneurism, but the simplest is always the best.
There are two classes, (1) dissecting and (2) circumscribed. (1) Dissecting aortic aneurism usually belongs to old age, and affects both sexes alike. It is
caused by weakening and rupture of the internal and
middle coats of the artery from fatty metamorphosis
due to senile decay. Inasmuch as aneurism is said to
be false when all the coats of the artery are not dilated
but some are ruptured or worn through, all dissecting
aortic aneurisms are necessarily also false. Dissecting
aneurism would also be said to be sacculated, fusiform,
and the like, according to the shape assumed.

(2) Circumscribed aortic aneurism is usually a man's
disease, and occurring generally at middle life or past.
Four fifths of the cases of thoracic, and about ninety
per cent. of abdominal circumscribed aneurism, occur
in men from forty to fifty years of age. This is due to
the difference from women in the mode of life and
occupation. About five per cent. of the cases occur
before thirty, and in all such cases observed by me
there has been a clear history of syphilis.

Circumscribed aneurism may be false or true, according as to whether or not some of the coats have
sustained solution of continuity from some cause. The

inner and middle coats usually give way in false aneurism, so that the sac is chiefly formed by the outer coat. But in case of wounds, the outer coat may yield, allowing the middle coat to protrude, giving rise to what is termed hernial false aneurism. If all the coats are ruptured, from wounds or disease, a diffuse or, better, consecutive aneurism may result. When due to injury they usually occur in the case of smaller vessels than the aorta, since in the latter case the patient would be likely to bleed to death before the aneurism could be formed. Wherever they occur they usually become circumscribed. Other varieties are fusiform, cylindrical, or globular. Generally, however, circumscribed aneurism is irregular in shape, causing it to be sacculated. And inasmuch as only the outer coat is often left to form the sac in such cases, circumscribed sacculated aneurism is also usually false. In the fusiform variety, or where there is slight and regular dilatation, the aneurism is not infrequently true.

Etiology.—Two classes of causes favor the production of aneurism: (1) increase of blood pressure, and (2) diminution of resisting power in the walls of the vessel. (1) Increase of blood pressure is caused by heavy lifting or straining, occupations necessitating long-continued effort, compensating hypertrophy of the left ventricle in aortic regurgitation, and intemperance. The course of the artery must also be taken into consideration. The pressure will not be so great at any given point in an artery whose course is straight as it would be when the artery is curved. The sharper the curve is the greater will be the pressure, and this is always directed against the periphery. This is remarkably well illustrated in cases of aortic aneurism. Of 880 cases collected by Sibson, 632 affected the arch, while only 71 occurred in the thoracic aorta, which is straight, as will be fully described presently. (2) The walls of the artery are weakened by surgical injuries, or by constitutional disease tending to produce arteritis, inherited or acquired. Of these causes, syphilis stands at the head and front. Lead poisoning probably comes next. Then follow gout, rheumatism, and renal disease. Sometimes predisposition to aneurism seems to be inherited, so that it will be handed down from parents to children for several generations.

Relative Frequency of Site.—Of 880 cases collected by Sibson, 87 were situated in the sinuses of Valsalva, 193 in the ascending portion of the arch, 140 in the ascending and transverse portion of the arch, 120 in the transverse portion of the arch, 20 in the transverse and descending portion of the arch, and 72 in the descending portion of the arch. That is to say, of 880 cases of aortic aneurism, 632 occurred in the arch alone. Of these, 420 cases occurred in the ascending portion, 140 in the transverse, and 72 in the descending portion of the arch. The thoracic aorta was affected in only 71 cases, and the abdominal aorta in 177.

The principal reason for this marked difference in the frequency of aneurism in the parts of the aorta, as just given, is owing to the course which the vessel takes. In the arch, of course, and especially the ascending and transverse portions, the blood pressure is much greater than where the vessel is straight. Not only that, but the arch is nearest to the heart to receive all its force. In the descending portion of the

arch the number of cases was only 72, and in the thoracic aorta, only 71, both on account of the straight
course of the vessel, and differing only by one case.
But when we come to the abdominal aorta the number
rises to 177. Here the artery is much more exposed
to injury than it is in the thoracic cavity. Moreover,
the abdominal aorta is subject to be bent on itself, or

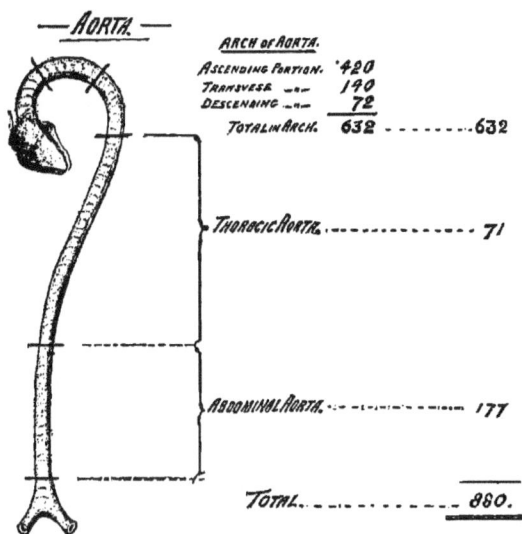

FIG. 31.—Schematic Diagram of Relative Frequency of Site of Aortic Aneurism.

put on the stretch, or twisted, and, in a word, to be
changed in its direction with every movement of which
the body is susceptible. In heavy lifting, wrestling,
the performances of athletes, and the effort at recover-
ing one's position when suddenly thrown off the balance by simple accident, all put a strain on the abdominal aorta, the habitual reception of which not
only tends to make that vessel brittle, but often is the
immediate cause of abdominal aneurism. Bartholow
18

states that he has never known a case of abdominal
aneurism that could not be directly traced to some act
of violence.

Symptoms.—In thoracic aneurism these may com-
mence suddenly, as if something had given way, but
much more frequently they come on gradually, with
failing health. There is pain, which is usually fixed,
but radiating. Pain is one of the first and most frequent symptoms of aortic aneurism in any part of the
vessel. It may and often does, exacerbate and remit,
but it is usually an early and a persistent symptom.
It is usually deep seated, extending through from
before back. Instead of actual pain it is sometimes
described by the patient as a feeling of soreness limited
to a small area. Besides pain, there are dyspnœa, with
more or less hoarse, stridulous cough, and alteration
of voice. The dyspnœa is of two kinds, (1) constant
and increasing and also (2) paroxysmal. The constant
and increasing dyspnœa of course is due to the grow-
ing aneurismal tumor pressing upon and displacing
important portions of the organs of respiration. The
paroxysmal dyspnœa, however, occurs in three ways:
(1) it may be due to spasm of the glottis, owing to irritation of the recurrent laryngeal nerves from pressure
of the aneurism; (2) paralytic closure of the glottis
from paralysis of these nerves from pressure of the
aneurismal tumor; and (3) pressure on the trachea with
accumulation of mucus at that point. Paralytic closure
of the glottis, due to pressure from the tumor, is necessarily a dangerous and often fatal symptom. The
greater the effort at inspiration, the more completely
are the walls of the larynx sucked together. Dys-

phagia from pressure on the œsophagus is not common, but headache due to obstruction to the return circulation of the blood is not infrequent. Sometimes there is disordered vision, owing to the change produced in the size of one or both pupils from pressure by the aneurismal tumor on the sympathetic nerves. One or both may be contracted or dilated, according as the sympathetic nerves are irritated or paralyzed. Slight hæmoptysis is of ordinary occurrence, the blood being mingled with the sputa. This slight hæmoptysis is due to bronchial congestion or pulmonary irritation, and is totally different from the rush of blood due to rupture of the sac. The patient gradually loses flesh and often has a careworn, wearied appearance. We see, therefore, that in thoracic aneurism, inward pressure signs, as they are termed, are always more or less prominent. In abdominal aneurism, on the other hand, with the exception of pain, which is present here as well as in thoracic aneurism, there are very few symptoms to be described by the patient. The onset of abdominal aneurism is, however, usually sudden. Indeed, Bartholow states that in all the cases of abdominal aneurism observed by him, the onset was sudden and definite, and traced to some act of violence, as sudden lifting of a heavy weight, wrestling, falling, or the like.

Physical Signs.—These differ according to the part of the aorta affected, the arch, the thoracic aorta, or the abdominal aorta. We will therefore consider them in their regular order.

I. The Arch of the Aorta.

Inspection.—At first, inspection may be purely negative. But after the aneurismal tumor has become sufficiently increased in size, a local bulging, or pulsating tumor, synchronous in its pulsations with the heart's systole, is usually observed at the right edge of the sternum in the second interspace, when the ascending portion of the arch is affected. This is by far the most common site for aneurism of the arch, since the ascending portion is most frequently affected. The tumor gradually increases downward to the right, pushing the apex of the heart downward and to the left. If the transverse portion of the arch is affected, the tumor may push forward the top of the sternum. Or it may appear on the left of the sternum, or at the base of the neck, according to the portion of the arch affected, and other circumstances. Pulsation may sometimes be noticed even in the interscapular region of the left side, if the descending portion of the arch be affected. The aneurismal tumor does not always pulsate. This occurs when the sac is filled to a great extent with fibrin, through which a small stream of blood flows, and especially if the descending portion of the arch be affected and the heart is weak. If the aneurism press on the superior vena cava, there will be enlargement of the veins on both sides of the neck, with more or less lividity of the face. But if the tumor press on one innominate vein only, enlargement of the veins and lividity of the face will be observed on the corresponding side only. The patient is not infrequently observed to have lost somewhat in flesh.

Palpation.—Two centres of pulsation, synchronous with systole, are usually felt, one due to the impulse of the heart, the other to the aneurism. The pulsation caused by the aneurism is usually accompanied by thrill, unless the sac is greatly filled with fibrin. In

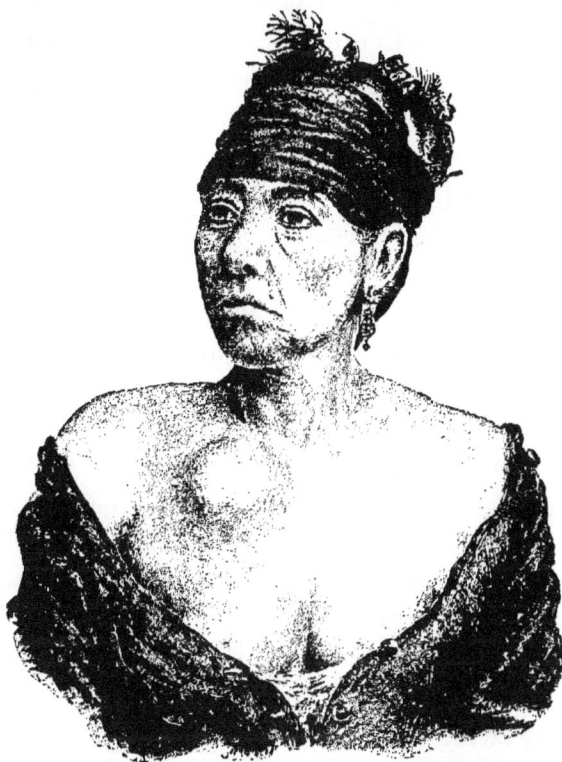

FIG. 32.—Aneurism of Ascending Portion of Arch in a German woman, æt. 48. Death from Rupture into Pericardial Sac.

that case, thrill may be, and usually is, entirely absent. Indeed, in these cases, also, the systolic impulse of the tumor may be so feeble that it can scarcely be observed.

In aneurism of the transverse portion of the arch the pulse is weaker at the left wrist, and on the left side of

the head and neck, than the right. This was beauti-
fully illustrated in a patient recently examined at my
request by Drs. Edward G. Janeway, Francis Delafield,
Alfred L. Loomis, and John A. Wyeth, of this city.
The patient was sent to Mt. Sinai Hospital, where Dr.
Wyeth ligated the left common carotid and left subclavian arteries with every prospect of success, but the
patient unfortunately died of pneumonia and syphilitic
pulmonary deposits. *Post-mortem* examination showed
the absolute accuracy of the diagnosis of aneurism of
the transverse portion of the arch extending to the junction of the descending portion. There was no dysphagia,
but the aneurismal sac, filled with fibrin, was adherent
to the trachea for about an inch and a half, causing
most distressing dyspnœa. In these cases, as also happens sometimes in aneurism of the descending portion
of the arch, the pulsation of the aneurism may be felt
in the suprasternal notch by pressing the finger well
down into it with the patient's head bent forward.
The vocal fremitus over the tumor is usually diminished or absent, according to the size of the aneurism
and displacement of lung tissue. Pressure on a large
bronchial tube also may so obstruct the convection of
the voice sound that the vocal fremitus may be entirely
absent over the corresponding area.

Percussion.—This should be gently performed over
the tumor, otherwise it causes great suffering to the
patient, to say nothing of the danger of rupture of the
aneurismal sac. For this reason auscultatory percussion is the best method, since it is performed very gently. Dullness is elicited over and immediately around
the tumor, and the dullness will be marked in propor

tion to the size and locality of the tumor. If it be
small and deep seated, the quality of the percussion
note may be very little changed. On the other hand,
if the tumor be large and superficial the quality may
be nearly or quite flat. It is of great importance to
observe whether the dullness extends continuously
out toward the acromial angle, over pulmonary tissue,
or across the median line. The latter sign would be a
sure indication of the presence of a tumor of some sort
in the mediastinum. If the descending or transverse
portions of the arch be affected, but especially the descending, dullness on percussion may be obtained in
the interscapular region of the left side.

Auscultation.—The aneurismal sounds usually present are most audible directly over the tumor. The
aneurismal systolic shock is usually accompanied by a
bruit, or murmur, which is louder than the heart sounds
and usually lower in pitch, especially when blowing in
quality. The bruit or murmur may, however, be rasping or filing in quality, and then the pitch may be high.
Sometimes it is roaring or whistling.

Besides the systolic bruit, there may be also a diastolic murmur, which is usually softer than the first, and
causing with it the to-and-fro sound. The aneurismal
bruit is usually heard in front, but it is sometimes even
heard posteriorly, in the interscapular space of the left
side, if the descending portion of the arch be affected.
There is diminution or absence of the respiratory murmur over the tumor, and pressure on a bronchial tube
may produce atelectasis for a corresponding area of
pulmonary tissue (see p. 60). Owing to the presence
usually of secondary localized bronchitis and a little

solidified lung tissue from pressure or local inflammation near the tumor, râles, bronchial breathing, bronchophony, and increased vocal fremitus over a corresponding small area, may be obtained.

Diagnosis.—To differentiate between aneurism of the three portions of the arch we must bear in mind, first, that the ascending portion is by far more frequently affected than the others, the transverse portion being next in order, and, lastly, the descending portion. When the ascending portion of the arch is the seat of the aneurism, the tumor, as already stated, usually appears in the second interspace at the right edge of the sternum, and gradually increases downward to the right, pushing the apex of the heart downward and to the left. Occurring in the transverse portion, it pushes the manubrium forward or appears at the left of the sternum.

Aneurism of the transverse portion of the arch causes a weaker pulse at the left wrist and on the left side of the head and neck than on the right, with pulsation in the suprasternal notch, and sometimes dullness on percussion, even in the interscapular region of the left side. Pressure on the trachea and œsophagus is more marked in these cases also. In case of aneurism of the descending portion of the arch of the aorta, there are pain in the interscapular region of the left side, dullness on percussion; and sometimes a pulsation is observed there, with a bruit on auscultation. Pulsation may also be sometimes felt in the suprasternal notch from aneurism of the descending portion of the arch, but not so distinctly as when the transverse portion is affected.

Arteria Innominata.—Aneurism of this vessel pulsates behind, or above, the inner part of the clavicle, causes weaker pulse on the right side than the left, and is rarely attended with dysphagia or tracheal pressure, but more frequently with pain or paralytic symptoms in the right arm. Pulsation of this aneurism diminishes or ceases from compression beyond the tumor.

Consolidation of pulmonary tissue from phthisis or syphilis, would give rise to dullness, which, however, would extend outward to the acromial angle, but not across the median line. There would also be wanting the inward pressure signs of aneurism. In the case of suspected pulmonary syphiloma, the failure of proper antisyphilitic treatment would rather favor the presence of aneurism.

Cancer of lungs may be associated with cancer of the mediastinal glands, which, becoming enlarged, would give rise to dullness that extended across the median line. But infiltrated cancer causes retraction of the chest walls instead of bulging, and there are no inward pressure signs. The cancerous cachexia and appearance of cancer elsewhere would establish the diagnosis.

Mediastinal tumors are the most difficult to differentiate. But unless associated with infiltrated cancer of the lungs or elsewhere, they usually occur in women under twenty-five, which aneurism rarely, if it ever does. Such tumors are usually also associated with currant-jelly (cancerous) expectoration, distention of the superficial veins on the chest, sometimes œdema of the chest and arm, and they may also exist elsewhere.

Coarctation and stricture of the aorta will give rise

to a systolic murmur, but they cause no bulging to be observed on inspection, no dullness on percussion, and no pressure signs. They usually result from syphilis, and coarctation is sometimes a congenital malformation.

Pulsating empyema is easily distinguished by the equality of the radial pulse, the absence of murmurs and thrill, as well as of tracheal, œsophageal, and laryngeal symptoms. It might occur to the practitioner to explore with a fine needle in order to set the question at rest, but this should not be done unless absolutely necessary, which is rarely the case, since emboli might be detached which would prove to be troublesome, if not fatal.

Pericardial effusion gives rise to prominence of the præcordial region, with more or less dyspnœa sometimes, and marked dullness on percussion; but the area of dullness is somewhat triangular, with the base down, aneurism, perhaps, never.

Subperiosteal abscess of the sternum may cause some prominence of the sternum, with dullness on percussion, but the inward pressure signs and all other signs of aneurism are wanting.

Cardiac hypertrophy causes only one centre of motion; when aneurism is present there are usually two. The aneurism may be situated, however, very close to the heart, and associated with aortic regurgitation and enlargement of the left ventricle. The absence of pressure signs and the presence of dropsy both favor cardiac disease. In aortic regurgitation, also, the pulse characteristic of that disease, and felt equally at both wrists, would be against aneurism.

II. Thoracic Aorta.

Aneurism of the thoracic aorta is not so easily recognized as when it occurs in the ascending and transverse portions of the arch. From a number of cases reported by Deputy-coroner Jenkins, of this city, and referred to by Dr. H. M. Biggs in a very interesting paper on this subject, read before the Section in Practice, New York Academy of Medicine, in February, 1888, it appears that not infrequently the cause of sudden death was due to rupture of unsuspected aneurism of the thoracic aorta. Owing to the position of the vessel, the physical signs are referable to the left side of the spinal column rather than the right, though an exception to this rule is rarely met with. Pain in this case, as elsewhere, is one of the symptoms, and usually consists of a gnawing sensation felt in the dorsal vertebræ. These may become eroded in time, and give rise to curvature of the spine. Bulging in a few cases may be noticed posteriorly, but dullness on percussion, over a circumscribed area, corresponding to the aneurism, is much more frequent. On auscultation a bruit may be heard, but is often absent. Owing to want of physical signs the aneurism often escapes detection, as already stated. Laryngeal symptoms are, of course, usually wanting, but there may be dysphagia from pressure on the œsophagus. The disease may be mistaken for pleurisy with effusion in some instances, so that the exploring needle alone could enable one to distinguish between them.

III. Abdominal Aorta.

The symptoms in this case have reference to pressure on abdominal organs. Pain, as in aneurism elsewhere, is one of the first symptoms. It may be local, or it may extend along the branches of the lumbar plexus. Jaundice from pressure on the bile duct is not common, but sometimes occurs. Changes in the urine from pressure on the renal vessels is even more rare. But nausea and vomiting are not infrequent, due to pressure against the stomach.

Inspection.—This is usually negative in its results, but in case of an emaciated patient, pulsation of the tumor may be visible in the recumbent dorsal position.

Palpation.—A pulsating tumor is usually felt somewhat to the left of the median line. The pulsation is synchronous with the cardiac systole, and is described as expansile in character—that is, it expands in all directions under the grasp of the hand. Thrill may also be present. Some authors describe this pulsation as post systolic, or coming just after the systole of the heart. Others regard it as purely systolic.

Percussion.—If the tumor is of considerable size there is dullness on percussion. But this is the least constant physical sign, owing to the presence of gas in the neighboring viscera.

Auscultation.—A systolic bruit may or may not be present here as elsewhere. If the tumor be well filled with fibrin there will be no bruit. Diastolic bruit is rare, but when present is thought to be diagnostic of the presence of aneurism.

Pulsation of the abdominal aorta may be mistaken

for abdominal aneurism. But in the former case the pulsation will be along the course of the vessel, giving, under palpation, the sense of a pulsating cord rather than an expansile tumor. The fact that such pulsations of the aorta usually occur in young and nervous women with thin abdominal walls, rather than in middle-aged men, also is against aneurism.

Pulsating tumors may also simulate aneurism and be even accompanied by a bruit. But by placing the patient in the knee-chest position the pulsation at once ceases if it be not aneurism, since the tumor simply gravitates away from the aorta and no longer has its pulsations imparted to it. In case of a young hysterical woman, recently examined by me at the Polyclinic, there was a distinct pulsating tumor felt over the abdominal aorta. The pulsation immediately ceased in the knee-chest position, and as she gave the history of constipation, I concluded that it was a case of impacted fæces. A dose of castor oil confirmed the diagnosis, by causing a large evacuation of the bowels and disappearance of the tumor. In this case I may add that the tumor had a distinctly boggy feeling, and was not expansile, but simply thumping under palpation.

THE END.

INDEX.